Essential
Student
Algebra
———

VOLUME ONE
———

Sets and Mappings

———

G000244641

Essential
Student
Algebra

VOLUME ONE

Sets and Mappings

T. S. BLYTH & E. F. ROBERTSON
University of St Andrews

London   New York
CHAPMAN AND HALL

First published in 1986 by
Chapman and Hall Ltd
11 New Fetter Lane, London EC4P 4EE
Published in the USA by
Chapman and Hall
29 West 35th Street, New York NY 10001

Printed in Great Britain by
J. W. Arrowsmith Ltd., Bristol

ISBN 0 412 27880 4

British Library Cataloguing in Publication Data
Blyth, T. S.
Essential student algebra.
Vol 1: Sets and mappings
1. Algebra
I. Title   II. Robertson, E. F.
512   QA155
ISBN 0-412-27880-4

Contents

Preface

If, as it is often said, mathematics is the queen of science then algebra is surely the jewel in her crown. In the course of its vast development over the last half-century, algebra has emerged as the subject in which one can observe pure mathematical reasoning at its best. Its elegance is matched only by the ever-increasing number of its applications to an extraordinarily wide range of topics in areas other than 'pure' mathematics.

Here our objective is to present, in the form of a series of five concise volumes, the fundamentals of the subject. Broadly speaking, we have covered in all the now traditional syllabus that is found in first and second year university courses, as well as some third year material. Further study would be at the level of 'honours options'. The reasoning that lies behind this modular presentation is simple, namely to allow the student (be he a mathematician or not) to read the subject in a way that is more appropriate to the length, content, and extent, of the various courses he has to take.

Although we have taken great pains to include a wide selection of illustrative examples, we have not included any exercises. For a suitable companion collection of worked examples, we would refer the reader to our series *Algebra through practice* (Cambridge University Press), the first five books of which are appropriate to the material covered here.

T.S.B., E.F.R.

CHAPTER ONE

Sets

Our approach to the theory of sets will be naïve in the sense that, rather than give a precise definition of the concept of a 'set', we shall rely heavily on intuition. By a *set* we shall mean a 'collection of objects', these objects being anything we care to imagine. The objects that make up a given set E are called the *elements* of E. We express the fact that an object x is an element of the set E by writing $x \in E$. If x is not an element of E then we write $x \notin E$. We say that two sets E, F are *equal*, and write $E = F$, if they contain precisely the same elements.

In depicting a set it is often convenient to simply list its elements. There is a standard notation for this : we enclose the list between 'curly brackets' or 'braces' as in

$$E = \{x, y, z, \ldots\}.$$

Example The set of positive divisors of 12 can be written

$$E = \{1, 2, 3, 4, 6, 12\}.$$

We have $6 \in E$ but $7 \notin E$.

Example The set of solutions of the equation $x^2 - 3x + 2 = 0$ can be written as $E = \{1, 2\}$.

Example $\{1, 2, 3, 4, 5\} = \{5, 4, 3, 2, 1\} = \{1, 3, 5, 2, 4\} =$ etc.

In dealing with many different sets a variant of the curly bracket notation is also very useful. We shall use the general notation

$$E = \{x \ ; \ \text{a statement } P(x) \text{ involving } x\}$$

to mean that 'E is the set of objects x for which the statement $P(x)$ is true'.

Example $\{1, 2, 3, 4, 6, 12\} = \{x \; ; \; x \text{ is a positive divisor of } 12\}$.

Example $\{1, 2\} = \{x \; ; \; x^2 - 3x + 2 = 0\}$.

On closer inspection, however, we see that this notation needs to be made less ambiguous. For example, $\{n \; ; \; 1 \le n \le 100\}$ could mean the (finite) set consisting of the first 100 positive integers, or the (infinite) set of all real numbers lying between 1 and 100. We obviate this difficulty by specifying a set from which the elements are to be chosen.

Example If $\mathbb{N} = \{0, 1, 2, 3, \ldots\}$ denotes the set of *natural numbers* then

$$E = \{n \in \mathbb{N} \; ; \; 1 \le n \le 100\}$$

is the collection consisting of the first 100 positive integers.

Example If $\mathbb{Z} = \{\ldots, -2, -1, 0, 1, 2, \ldots\}$ is the set of *integers* then

$$F = \{n \in \mathbb{Z} \; ; \; 1 \le n \le 100\}$$

also describes the collection consisting of the first 100 positive integers.

Example If $\mathbb{Q} = \{\frac{m}{n} \; ; \; m, n \in \mathbb{Z}, n \ne 0\}$ is the set of *rational numbers* then

$$G = \{x \in \mathbb{Q} \; ; \; 1 \le x \le 2\}$$

describes the set of rational numbers lying between 1 and 2.

Example If $\mathbb{R}$ is the set of *real numbers* then

$$H = \{x \in \mathbb{R} \; ; \; 1 \le x \le 2\}$$

describes the set of real numbers lying between 1 and 2. More generally, if $a, b \in \mathbb{R}$ with $a \le b$ then the *closed interval* $[a, b]$ is defined to be

$$[a, b] = \{x \in \mathbb{R} \; ; \; a \le x \le b\},$$

and the *open interval* $]a, b[$ is defined to be

$$]a, b[= \{x \in \mathbb{R} \; ; \; a < x < b\}.$$

We can also define the *half-open* or *half-closed* intervals

$$]a, b] = \{x \in \mathbb{R} \; ; \; a < x \leq b\};$$
$$[a, b[= \{x \in \mathbb{R} \; ; \; a \leq x < b\}.$$

Example If $\mathbb{C} = \{a + ib \; ; \; a, b \in \mathbb{R}, \; i^2 = -1\}$ is the set of *complex numbers* then

$$I = \{z \in \mathbb{C} \; ; \; z^n = 1\}$$

describes the set of complex numbers lying on the unit circle in the Argand diagram.

Example Consider the set described by

$$X = \{x \in E \; ; \; (x + 5)(x - 2)(3x - 7)(x^2 - 2)(x^2 + 1) = 0\},$$

where E is one of $\mathbb{N}, \mathbb{Z}, \mathbb{Q}, \mathbb{R}, \mathbb{C}$. The following table describes X in each case :

E	X
$\mathbb{N}$	$\{2\}$
$\mathbb{Z}$	$\{-5, 2\}$
$\mathbb{Q}$	$\{-5, 2, \frac{7}{3}\}$
$\mathbb{R}$	$\{-5, 2, \frac{7}{3}, \sqrt{2}, -\sqrt{2}\}$
$\mathbb{C}$	$\{-5, 2, \frac{7}{3}, \sqrt{2}, -\sqrt{2}, i, -i\}$

By omitting some of the elements in a set we obtain other sets called subsets. More formally, by a *subset* of a set E we mean a set F such that every element of F is an element of E. Using the logical symbol $\Rightarrow$ to mean 'implies' we can indicate that F is a subset of E by writing

$$x \in F \Longrightarrow x \in E.$$

We shall denote the fact that F is a subset of E by writing $F \subseteq E$. Sometimes we write this in the equivalent form $E \supseteq F$. If $F \subseteq E$ and $F \neq E$ then we shall write $F \subset E$ and say that F is a *proper subset* of E.

Example $\mathbb{N} \subset \mathbb{Z} \subset \mathbb{Q} \subset \mathbb{R} \subset \mathbb{C}$.

1.1 Theorem *$E = F$ if and only if $E \subseteq F$ and $F \subseteq E$.*

Proof If $E = F$ then clearly we have both $E \subseteq F$ and $F \subseteq E$. Conversely, suppose that $E \subseteq F$ and $F \subseteq E$. Then every element of E is an element of F, and every element of F is an element of E. Consequently E and F have the same elements and so are equal. $\diamond$

Using the logical symbol $\leftrightarrow$ to mean 'implies and is implied by', or more commonly 'if and only if', we can indicate that $E = F$ by writing

$$x \in E \iff x \in F.$$

Example $\{n \in \mathbb{N} \; ; \; n \text{ is even}\} = \{n \in \mathbb{N} \; ; \; n^2 \text{ is even}\}$. In fact, let the left-hand side be the set A and the right-hand side the set B. Then $A \subseteq B$ since

$$n \in A \implies n = 2m \implies n^2 = 4m^2 \implies n \in B;$$

and $B \subseteq A$ since

$$n \notin A \implies n = 2m + 1 \implies n^2 = 4m^2 + 4m + 1 \implies n \notin B,$$

which is logically equivalent to the statement

$$n \in B \implies n \in A.$$

1.2 Theorem *If $E \subseteq F$ and $F \subseteq G$ then $E \subseteq G$.*

Proof Every element of E is an element of F, and every element of F is an element of G. Thus every element of E is an element of G. $\diamond$

We now extend our intuitive idea of a set to allow as a set a 'collection consisting of no objects at all'. The courtesy of regarding this as a set has several advantages. This conventional *empty set* is generally denoted by the Norwegian letter $\emptyset$. In allowing $\emptyset$ the status of a set, we gain the advantage of being able to talk about a set without knowing whether or not it has any elements. A set E that does have elements will be called *non-empty*, and we shall indicate this by writing $E \neq \emptyset$.

Consider now, for any object x, the statement

$$x \in E \text{ and } x \notin E.$$

By the logical principle of non-contradiction, this statement is false. Roughly speaking, therefore, there is no object x such that $x \in E$ and $x \notin E$. The concept of an empty set usefully describes this situation; for every set E we can write

$$\emptyset = \{x \; ; \; x \in E \text{ and } x \notin E\}.$$

For every set E we have $\emptyset \subseteq E$ since the statement 'every element of $\emptyset$ is an element of E' can be regarded as true simply because $\emptyset$ has no elements at all. When E is non-empty we have $\emptyset \subset E$.

It is sometimes convenient to represent sets by means of *Venn diagrams*, the interpretation of which is as follows. Since we assume that a given object either belongs to a given set or does not, we can represent a set by the interior of a closed contour and if an object belongs to the set then we represent it by a point inside the contour, whereas if it does not belong to the set then we represent it by a point outside the contour. Thus, for example,

indicates that $x \in E$ and $y \notin E$. This is also useful in exhibiting subsets. We can exhibit $E \subseteq F$ by the diagram

Given sets E and F, we define their *intersection* to be the set, denoted by $E \cap F$, consisting of those elements that are in both E and F. Thus

$$E \cap F = \{x \; ; \; x \in E \text{ and } x \in F\},$$

and we can depict this by the Venn diagram

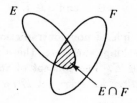

The principal properties of intersection are the following.

1.3 Theorem *If A, B, C are sets then*

(1) $A \cap \emptyset = \emptyset$;
(2) $A \cap A = A$;
(3) $A \cap B = B \cap A$;
(4) $(A \cap B) \cap C = A \cap (B \cap C)$.

Proof The proof is immediate in each case. ◇

Note that (3) is often referred to by saying that $\cap$ is *commutative*, and that (4) is referred to by saying that $\cap$ is *associative*. We can, and often do, represent each side of (4) by writing simply $A \cap B \cap C$.

We define the *union* of sets E, F to be the set, denoted by $E \cup F$, that consists of those objects that are in E or in F (or in both). Thus

$$E \cup F = \{x \; ; \; x \in E \text{ or } x \in F\},$$

and we can depict this by the Venn diagram

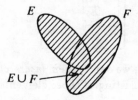

The principal properties of union are the following.

1.4 Theorem *If A, B, C are sets then*

(1) $A \cup \emptyset = A$;
(2) $A \cup A = A$;
(3) $A \cup B = B \cup A$;
(4) $(A \cup B) \cup C = A \cup (B \cup C)$.

Proof Again, this is immediate. $\diamond$

By (3) and (4), the operation of union is also commutative and associative, and we often express each side of (4) by writing simply $A \cup B \cup C$.

Example For every positive integer n let $n\mathbb{Z} = \{nk \; ; \; k \in \mathbb{Z}\}$ be the set of all integer multiples of n. Then $p\mathbb{Z} \subseteq n\mathbb{Z}$ if and only if n divides p, so $n\mathbb{Z} \cap m\mathbb{Z} = p\mathbb{Z}$ where $p = \mathrm{lcm}\{n, m\}$. Thus, for example,

$$(3\mathbb{Z} \cap 6\mathbb{Z}) \cup 12\mathbb{Z} = 6\mathbb{Z} \cup 12\mathbb{Z} = 6\mathbb{Z}.$$

The following *absorption laws* are useful.

1.5 Theorem *If A, B are sets then*

(1) $A \cap (A \cup B) = A$;
(2) $A \cup (A \cap B) = A$.

Proof These follow immediately from the inclusions $A \subseteq A \cup B$ and $A \cap B \subseteq A$. $\diamond$

Unions and intersections are also linked by the following *distributive laws*.

1.6 Theorem *If A, B are sets then*

(1) $A \cap (B \cup C) = (A \cap B) \cup (A \cap C)$;
(2) $A \cup (B \cap C) = (A \cup B) \cap (A \cup C)$.

Proof We prove (1), the proof of (2) being similar. By 1.1, it suffices to establish the inclusions

$$A \cap (B \cup C) \subseteq (A \cap B) \cup (A \cap C),$$
$$A \cap (B \cup C) \supseteq (A \cap B) \cup (A \cap C).$$

To establish the first inclusion, observe that if $x \in A \cap (B \cup C)$ the $x \in A$ and $x \in B \cup C$, so $x \in A$ and either $x \in B$ or

$x \in C$. Thus $x \in A$ and $x \in B$, or $x \in A$ and $x \in C$, whence $x \in (A \cap B) \cup (A \cap C)$.

As for the second, suppose that

$$x \in (A \cap B) \cup (A \cap C).$$

Then either $x \in A \cap B$ or $x \in A \cap C$ (or both). So either $x \in A$ and $x \in B$, or $x \in A$ and $x \in C$. Clearly then we have $x \in A$, and either $x \in B$ or $x \in C$. In other words, we have

$$x \in A \cap (B \cup C)$$

and the result follows. $\diamond$

1.7 Corollary *If $A \cap B = A \cap C$ and $A \cup B = A \cup C$ then $B = C$.*

Proof If the conditions hold then, by 1.5 and 1.6, we have

$$\begin{aligned}
B = B \cap (B \cup A) &= B \cap (C \cup A) \\
&= (B \cap C) \cup (B \cap A) \\
&= (B \cap C) \cup (C \cap A) \\
&= C \cap (B \cup A) \\
&= C \cap (C \cup A) = C. \quad \diamond
\end{aligned}$$

Example If A, B, C are sets then we have the equality

$$(A \cap B) \cup (B \cap C) \cup (C \cap A) = (A \cup B) \cap (B \cup C) \cap (C \cup A).$$

To establish this, we again use 1.5 and 1.6 :

$$\begin{aligned}
(A \cup B) \cap (B \cup C) \cap (C \cup A) &= [A \cup (B \cap C)] \cap (B \cup C) \\
&= [A \cap (B \cup C)] \cup [B \cap C \cap (B \cup C)] \\
&= (A \cap B) \cup (A \cap C) \cup (B \cap C).
\end{aligned}$$

This equality can also be verified using Venn diagrams.

If E is a set and A is a subset of E then we define the *complement of A in E* to be the subset consisting of those elements of E that do not belong to A. This set is variously denoted by $C_E(A), A^c, E - A, E \setminus A$, or simply A'. When no confusion can arise over the set E we shall use the notation A'; otherwise we shall use the notation $C_E(A)$. Thus we have

$$C_E(A) = A' = \{x \in E \ ; \ x \notin A\}.$$

The corresponding Venn diagram is

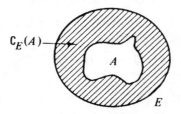

It is clear from the above definition that for every subset A of E we have

$$A \cap A' = \emptyset \quad \text{and} \quad A \cup A' = E.$$

We shall now show that these properties characterise complements.

1.8 Theorem *Let A and X be subsets of E. Then*

(1) *$A \cap X = \emptyset$ only if $X \subseteq A'$;*
(2) *$A \cup X = E$ only if $X \supseteq A'$.*

Proof (1) If $A \cap X = \emptyset$ then taking the union of each side with A' we obtain

$$(A \cap X) \cup A' = \emptyset \cup A' = A'.$$

But by distributivity the left-hand side is

$$(A \cup A') \cap (X \cup A') = E \cap (X \cup A') = X \cup A' \supseteq X.$$

The proof of (2) is similar. $\diamond$

1.9 Corollary *If $A \cap X = \emptyset$ and $A \cup X = E$ then $X = A'$.* $\diamondsuit$

1.10 Corollary $(A')' = A$. $\diamondsuit$

The following *de Morgan laws* are important.

1.11 Theorem *If A, B are subsets of E then*

(1) $(A \cap B)' = A' \cup B'$;

(2) $(A \cup B)' = A' \cap B'$.

Proof (1) It suffices to observe that

$$A \cap B \cap (A' \cup B') = (A \cap B \cap A') \cup (A \cap B \cap B') = \emptyset;$$
$$(A \cap B) \cup A' \cup B' = (A \cup A' \cup B') \cap (B \cup A' \cup B') = E,$$

and hence that the complement of $A \cap B$ is $A' \cup B'$.

The proof of (2) is similar. $\diamondsuit$

1.12 Corollary $A \subseteq B \iff B' \subseteq A'$.

Proof By 1.8 we have

$$A \subseteq B \iff A \cap B = A$$
$$\iff A' \cup B' = (A \cap B)' = A'$$
$$\iff B' \subseteq A'. \quad \diamondsuit$$

If A, B are subsets of a set E then we define the *difference set* $A \setminus B$ (some write $A - B$) by

$$A \setminus B = \{x \in E \ ; \ x \in A, x \notin B\}.$$

The corresponding Venn diagram is

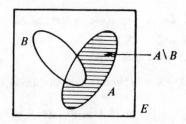

It is clear that $A \setminus B = A \cap B'$.

For a given finite set E we shall use the notation $|E|$ to denote the number of elements in E. If A, B are finite sets that are *disjoint*, in the sense that $A \cap B = \emptyset$, then it is clear that $|A \cup B| = |A| + |B|$.

1.13 Theorem *If A, B are finite sets then*

$$|A \cup B| = |A| + |B| - |A \cap B|.$$

Proof The preceding Venn diagram shows that $A \setminus B, A \cap B$, and $B \setminus A$ are pairwise disjoint and make up $A \cup B$. Consequently,

$$|A \cup B| = |A \setminus B| + |A \cap B| + |B \setminus A|.$$

But $A \cap B$ and $A \setminus B$ are disjoint and make up A, so

$$|A| = |A \cap B| + |A \setminus B|.$$

Similarly, we have $|B| = |A \cap B| + |B \setminus A|$, and therefore

$$|A \cup B| = |A| - |A \cap B| + |A \cap B| + |B| - |A \cap B|$$
$$= |A| + |B| - |A \cap B|. \quad \Diamond$$

1.14 Corollary *If A, B, C are finite sets then*

$$|A \cup B \cup C| = |A| + |B| + |C| - |A \cap B| - |B \cap C| - |C \cap A| + |A \cap B \cap C|.$$

Proof By 1.13, we have

$$|A \cup B \cup C| = |A \cup (B \cup C)|$$
$$= |A| + |B \cup C| - |A \cap (B \cup C)|$$
$$= |A| + |B| + |C| - |B \cap C| - |(A \cap B) \cup (A \cap C)|$$
$$= |A| + |B| + |C| - |B \cap C| - |A \cap B| - |A \cap C| + |A \cap B \cap C|$$

as required. $\Diamond$

Example At an international meeting of 100 English, French, and German mathematicians, at least 75 speak English, at least 70 speak French, and at least 65 speak German. At least how many speak all three languages?

This type of problem can be solved by using the above results. Let E, F, G be the sets of English, French, German speakers. Then, by 1.13,

$$|E \cap F| = |E| + |F| - |E \cup F| \geq 75 + 70 - 100 = 45,$$
$$|E \cap G| = |E| + |G| - |E \cup G| \geq 75 + 65 - 100 = 40,$$
$$|F \cap G| = |F| + |G| - |F \cup G| \geq 70 + 65 - 100 = 35.$$

Again using 1.13, we then see that

$$|E \cap F \cap G| = |E \cap (F \cap G)|$$
$$= |E| + |F \cap G| - |E \cup (F \cap G)|$$
$$\geq 75 + 35 - 100 = 10.$$

A Venn diagram illustrating this minimum number is

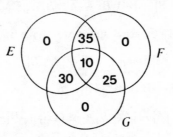

If we are given the further information that one participant of each nationality cannot converse other than in his native tongue, then we must have

$$|E \cup F|, |E \cup G|, |F \cup G| \leq 99$$

in which case

$$|E \cap F| \geq 46, \quad |E \cap G| \geq 41, \quad |F \cap G| \geq 36.$$

Since 1 speaks only French and 1 speaks only German, we also have

$$|E \cup (F \cap G)| \leq 98.$$

Consequently we see that

$$|E \cap F \cap G| \geq 75 + 36 - 98 = 13.$$

The Venn diagram

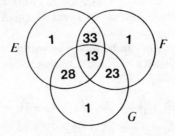

illustrates the minimum number in this case.

Very often in mathematics one has to deal with a *set of sets*, i.e. a set whose elements are themselves sets. For example, given a set E the set that consists of all the subsets of E is called the *power set* of E and will be denoted by $\mathbf{P}(E)$.

Example If $E = \{x, y\}$ then $\mathbf{P}(E) = \{\emptyset, \{x\}, \{y\}, E\}$.

Example If $E = \{x, y, z\}$ then

$$\mathbf{P}(E) = \{\emptyset, \{x\}, \{y\}, \{z\}, \{x, y\}, \{x, z\}, \{y, z\}, E\}.$$

A set that consists of a single element is called a *singleton*. Note that for every object x we have $x \in \{x\}$ and $x \neq \{x\}$. In particular, we have $\emptyset \in \{\emptyset\}$ and $\emptyset \neq \{\emptyset\}$. The error of confusing $\emptyset$ with $\{\emptyset\}$ can be avoided by remembering that 'a bag that contains an empty bag is not empty'.

Example If $E = \{1, 2, \{1, 2\}\}$ then, as in the preceding Example, $\mathbf{P}(E)$ has eight elements, namely

$$\emptyset, \{1\}, \{2\}, \{\{1, 2\}\}, \{1, 2\}, \{1, \{1, 2\}\}, \{2, \{1, 2\}\}, E.$$

Observe that $E \cap \mathbf{P}(E) = \{\{1, 2\}\}$ since $\{1, 2\}$ is the only element of E that is also an element of $\mathbf{P}(E)$.

It should be noted carefully that, despite one's intuitive inclination to reject the possibility, it is possible to have both $x \in A$ and $x \subseteq A$; in other words, for an object to be both an element and a subset of a set.

Example If $E = \{1, 2, \{1, 2\}\}$ then $x = \{1, 2\}$ is such that $x \in E$ and $x \subseteq E$.

Example An infinite set every element of which is a subset is

$$\{\emptyset, \{\emptyset\}, \{\emptyset, \{\emptyset\}\}, \{\emptyset, \{\emptyset\}, \{\{\emptyset\}\}\}, \ldots\}.$$

Definition Given objects x, y we define the *ordered pair* (x, y) by

$$(x, y) = \{\{x\}, \{x, y\}\}.$$

1.15 Theorem *The ordered pairs (x, y) and (x', y') are equal if and only if $x = x'$ and $y = y'$.*

Proof Suppose first that $y \neq x$ and $y' \neq x'$. Then we have

$$(x, y) = (x', y') \iff \{\{x\}, \{x, y\}\} = \{\{x'\}, \{x', y'\}\}$$
$$\iff \{x\} = \{x'\} \text{ and } \{x, y\} = \{x', y'\}$$
$$\iff x = x' \text{ and } y = y'.$$

Suppose now that $y = x$. Then

$$(x, y) = (x, x) = \{\{x\}, \{x, x\}\} = \{\{x\}, \{x\}\} = \{\{x\}\}$$

and so in this case we have

$$(x, y) = (x', y') \iff \{\{x\}\} = \{\{x'\}, \{x', y'\}\}$$
$$\iff \{x\} = \{x'\} = \{x', y'\}$$
$$\iff x' = y' = x \, [= y].$$

The case where $y' = x'$ is similar. $\Diamond$

1.16 Corollary *If $x \neq y$ then $(x, y) \neq (y, x)$.* $\Diamond$

It is worth noting at this juncture that what is important about an ordered pair is not so much the way in which it is defined but rather the result of 1.15, which says that ordered pairs are equal if and only if corresponding components are the same.

Definition If E, F are sets then we define the *cartesian product set $E \times F$* by

$$E \times F = \{(x, y) \; ; \; x \in E, y \in F\}.$$

A pictorial representation of $E \times F$ is the following.

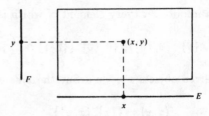

The notion of a cartesian product set has a geometric motivation, as the above pictorial representation begins to indicate.

Example $\mathbb{R} \times \mathbb{R}$ is the set of all ordered pairs of real numbers, i.e. represents the plane in cartesian geometry.

Example Plane curves can be regarded as subsets of $\mathbb{R} \times \mathbb{R}$. For example,

$$\{(x,y) \; ; \; x^2 + y^2 = 1\}$$

describes the unit circle, i.e. the circle of radius 1 centred at the origin.

Example Regions in the plane can often be similarly described. For example,

$$\{(x,y) \in \mathbb{R} \times \mathbb{R} \; ; \; x^2 + y^2 \leq 1\}$$

describes the unit disc.

The notion of an *ordered triple* (x, y, z) can usefully be defined as an ordered pair :

$$(x,y,z) = \big((x,y),z\big).$$

Using 1.15, we can see easily that $(x, y, z) = (x', y', z')$ if and only if $x = x', y = y', z = z'$. Proceeding in this manner, we can define for every positive integer n the notion of an *ordered n-tuple* $(x_1, x_2, \ldots, x_n)$. If $E_1, \ldots, E_n$ are sets then we define the cartesian product set $\bigtimes_{i=1}^{n} E_i = E_1 \times \cdots \times E_n$ to be the set of all ordered n-tuples $(x_1, \ldots, x_n)$ with $x_i \in E_i$ for $i = 1, \ldots, n$. In the case where all the E_i are the same set E we write this cartesian product set as E^n. Thus in particular $\mathbb{R}^2 = \mathbb{R} \times \mathbb{R}$ represents the cartesian plane, and $\mathbb{R}^3 = \mathbb{R} \times \mathbb{R} \times \mathbb{R}$ represents three-dimensional space.

Example $\{(x, y, z) \in \mathbb{R}^3 \; ; \; z = 4 - x^2 - y^2\}$ represents a paraboloid of revolution with apex at the point $(0, 0, 4)$. To see this, begin by fixing z, say $z = k$. Then the cross-section in the plane $z = k$ is

$$\{(x, y, k) \; ; \; x^2 + y^2 = 4 - k\}$$

which is a circle if $k \leq 4$ and is $\emptyset$ otherwise. Similarly, the cross-sections by the planes $x = 0$ and $y = 0$ are parabolae.

Example If $A = \{(x, y, 0) \; ; \; x^2 + y^2 = 1\}$ is the unit circle in the (x, y)-plane of $\mathbb{R}^3$ and if B is the z-axis of $\mathbb{R}^3$ then $A \times B$ is a circular cylinder.

CHAPTER TWO

Mappings

By a *relation* between sets A and B we shall mean intuitively a sentence $S(x, y)$ from which, on substituting for x an object of A and for y an object of B, we obtain a meaningful sentence that can be classified as true or false. The subset G of $A \times B$ consisting of those pairs $(a, b) \in A \times B$ for which $S(a, b)$ is true will be written

$$G = \{(a, b) \in A \times B \; ; \; S(a, b)\}$$

and called the *graph* of the relation.

Example Let $A = \{1, 2, 3, 4\}$ and $B = \{1, 2, 3\}$. Consider the relation $S(x, y)$ given by

$$x + y \le 4.$$

The graph of this, namely

$$G = \{(a, b) \in A \times B \; ; \; a + b \le 4\},$$

is the set of pairs marked • in the diagram

```
3 ○    •  ○  ○  ○
2 ○    •  •  ○  ○
1 ○    •  •  •  ○
B
       ○  ○  ○  ○  A
       1  2  3  4
```

Example If $A = B = \mathbb{R}$ the relation $x^2 + y^2 \le 1$ has as graph the unit disc.

Example If $A = B = \mathbb{Z}$ the relation $x^2 + y^2 = 1$ has as graph the set

$$\{(0, 1), (0, -1), (1, 0), (-1, 0)\}.$$

Example If $A = B = \mathbb{R}$ the relation $y = \sin x$ has as graph the sine curve.

Example If $A = B = \mathbb{R}$ then the graph of the relation

$$-1 \le x + y \le 1$$

is that part of the plane contained between the line $x + y = -1$ and the line $x + y = 1$:

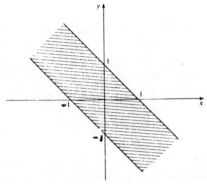

Very often we shall define a relation R between A and B by specifying its graph. More precisely, we shall say that $a \in A$ is *R-related* to $b \in B$, and write aRb, when (a, b) belongs to the graph of R. Thus we have

$$aRb \iff (a, b) \in G,$$

so we can regard G as the set of pairs that 'satisfy' the relation. Because of this, it is common practice to identify a relation between A and B with a subset of $A \times B$.

Using the logical symbol $\exists$ (the *existential quantifier*) to mean 'there exists', we define the *domain* of a relation R by

$$\text{Dom}\, R = \{a \in A\ ;\ (\exists b \in B)\ aRb\},$$

and the *image* of R by

$$\text{Im}\, R = \{b \in B\ ;\ (\exists a \in A)\ aRb\}.$$

These sets can be depicted pictorially as follows.

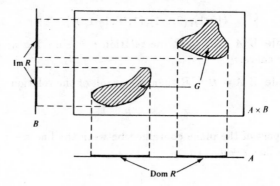

Example In the first of the above Examples the domain is $\{1,2,3\} \subset A$, and the image is B; in the second, the domain is the interval $[-1,1]$ on the x-axis, and the image is the interval $[-1,1]$ on the y-axis; in the third, the domain is $\{-1,1\}$ on the x-axis, and the image is $\{-1,1\}$ on the y-axis; in the fourth, the domain is the x-axis, and the image is the interval $[-1,1]$ on the y-axis; and in the fifth the domain is the x-axis and the image is the y-axis.

A particularly important type of relation, and what is probably the most important concept in the whole of mathematics, is the following.

Definition Given sets E and F, we define a *mapping* from E to F to be a relation R between E and F which is such that for every $x \in E$ the set $\{y \in F \; ; \; xRy\}$ is a singleton.

Example The relation S between $\mathbb{R}$ and $\mathbb{R}$ given by

$$xSy \iff y = \sin x$$

is a mapping from $\mathbb{R}$ to $\mathbb{R}$, there being only one y such that xSy, namely $\sin x$.

Note from this definition that if R is a mapping from E to F then the domain of R is the whole of E.

There is a universally accepted notation for mappings. We describe a mapping R from E to F by writing $R : E \rightarrow F$. This is more is traditionally written as $f : E \rightarrow F$. The unique occupant y of the singleton set $\{y \in F \; ; \; xRy\}$ is denoted by $f(x)$ and is called the *image of x under f*. With this notation, the graph G of the mapping f consists of all pairs of the form $\big(x, f(x)\big)$ and can be described by $G = \{(x, y) \in E \times F \; ; \; y = f(x)\}$. In writing $f : E \to F$ we sometimes call E the *departure set* of f and F the *arrival set* of f. Observe that the domain and the departure set are the same, whereas in general the image is a subset of the arrival set, namely

$$\operatorname{Im} f = \{y \in F \; ; \; (\exists x \in E) \; y = f(x)\} = \{f(x) \; ; \; x \in E\}.$$

As for the graph of f, this can be depicted in the following typical way which is consistent with similar diagrams used in calculus, where mappings are also called *functions*.

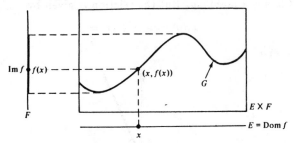

When no confusion can arise over the sets E and F, we shall often describe a mapping $f : E \to F$ by writing $x \mapsto f(x)$.

Example For every $x \in \mathbb{R}$ define

$$|x| = \begin{cases} x & \text{if } x \geq 0; \\ -x & \text{if } x < 0. \end{cases}$$

Then $x \mapsto |x|$ describes a mapping from $\mathbb{R}$ to $\mathbb{R}$. Its graph is

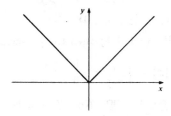

Example Consider the relation between $\mathbb{R}$ and $\mathbb{R}$ given by

$$xRy \iff x + |x| = y + |y|.$$

Observe that
$$x + |x| = \begin{cases} 2x & \text{if } x \geq 0; \\ 0 & \text{if } x < 0. \end{cases}$$
The graph of R is

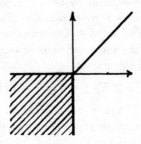

so R is not a mapping. But the relation S given by
$$xSy \iff x + |x| = y$$
has graph

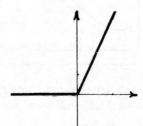

and is a mapping (function) $f : \mathbb{R} \to \mathbb{R}$. It can be described by $x \mapsto f(x)$ where
$$f(x) = \begin{cases} 0 & \text{if } x \leq 0; \\ 2x & \text{if } x > 0. \end{cases}$$
Clearly, we have $\text{Im } f = \{y \; ; \; y \geq 0\}$.

Example If $A = \mathbb{R} \setminus \{-1, 0, 1\}$ then the prescription
$$f(x) = \begin{cases} \frac{1}{x} & \text{if } |x| < 1; \\ -\frac{1}{x} & \text{if } |x| > 1, \end{cases}$$
defines a mapping $f : A \to \mathbb{R}$. Now if $B = \mathbb{R} \setminus \{0\}$ then the mapping $\vartheta : B \to \mathbb{R}$ described by
$$x \mapsto \vartheta(x) = \frac{1}{x}$$

has as graph the rectangular hyperbola, so the graph of f is readily seen to be

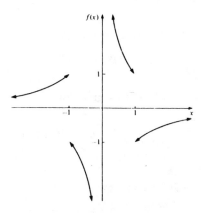

and Im f is the y-axis minus the set $\{-1, 0, 1\}$, i.e. Im $f = A$.

Consider now the diagram of sets and mappings

$$A \xrightarrow{\ f\ } B \xrightarrow{\ g\ } C,$$

passage through which can be indicated by

$$x \longmapsto f(x) \longmapsto g[f(x)].$$

Clearly, the assignment $x \mapsto g[f(x)]$ describes a mapping from A to C. We call this the *composite* of f and g and denote it by $g \circ f$. Thus $g \circ f : A \to C$ is given by the prescription

$$(g \circ f)(x) = g[f(x)].$$

Concerning this definition, the following points should be noted carefully :

(1) $g \circ f$ is defined only when the arrival set of f coincides with the departure set of g.

(2) If $g \circ f$ is defined then $f \circ g$ need not be. Indeed, for both to be defined we require $f : A \to B$ and $g : B \to A$.

(3) Even when $g \circ f$ and $f \circ g$ are defined they need not be equal. For example, when $f : A \to B$ and $g : B \to A$ we have $\mathrm{Dom}(g \circ f) = A$ and $\mathrm{Dom}(f \circ g) = B$, so if $A \neq B$ then $g \circ f \neq f \circ g$.

Example If $f : \mathbb{R} \to \mathbb{R}$ and $g : \mathbb{R} \to \mathbb{R}$ are given by $f(x) = |x^3|$ and $g(x) = e^x$ then the composite mappings $g \circ f$ and $f \circ g$ are given by the prescriptions

$$(g \circ f)(x) = e^{|x^3|},$$
$$(f \circ g)(x) = e^{3x}.$$

Example Let $f : \mathbb{R} \to \mathbb{R}$ be given by

$$f(x) = \begin{cases} 1 - x & \text{if } x \geq 0; \\ x^2 & \text{if } x < 0, \end{cases}$$

and let $g : \mathbb{R} \to \mathbb{R}$ be given by

$$g(x) = \begin{cases} x & \text{if } x \geq 0; \\ x - 1 & \text{if } x < 0. \end{cases}$$

Now we have

$$f[g(x)] = \begin{cases} f(x) & \text{if } x \geq 0; \\ f(x - 1) & \text{if } x < 0, \end{cases}$$
$$= \begin{cases} 1 - x & \text{if } x \geq 0; \\ (x - 1)^2 & \text{if } x < 0, \end{cases}$$

so the graph of $f \circ g$ is

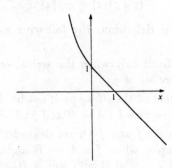

As for $g \circ f$, we have

$$g[f(x)] = \begin{cases} g(1-x) & \text{if } x \geq 0; \\ g(x^2) & \text{if } x < 0, \end{cases}$$

$$= \begin{cases} -x & \text{if } x > 1; \\ 1-x & \text{if } 0 \leq x \leq 1; \\ x^2 & \text{if } x < 0, \end{cases}$$

so the graph of $g \circ f$ is

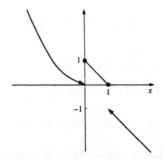

An important property of the operation of composition of mappings is that it is *associative*, in the following sense.

2.1 Theorem *Given* $A \xrightarrow{\ f\ } B \xrightarrow{\ g\ } C \xrightarrow{\ h\ } D$ *we have*

$$h \circ (g \circ f) = (h \circ g) \circ f.$$

Proof Clearly, each of the triple composites is defined and has departure set A and arrival set D. Now for every $x \in A$ we have

$$[h \circ (g \circ f)](x) = h[(g \circ f)(x)] = h[g(f(x))];$$
$$[(h \circ g) \circ f](x) = (h \circ g)[f(x)] = h[g(f(x))].$$

It follows that $h \circ (g \circ f)$ and $(h \circ g) \circ f$ have the same graphs. Consequently, they are equal. ◇

2.2 Corollary *If* $h \circ g \circ f : A \to D$ *is defined by* $(h \circ g \circ f)(x) = h[g(f(x))]$ *then*

$$h \circ (g \circ f) = (h \circ g) \circ f = h \circ g \circ f. \quad ◇$$

Example It is 2.2 that allows us to write, without excessive use of brackets, such expressions as, for example, $\log|\tan \vartheta|$.

Definition By the *identity map* on a set A we mean the mapping $\mathrm{id}_A : A \to A$ given by $\mathrm{id}_A(x) = x$ for every $x \in A$.

Clearly, the graph of id_A is the diagonal of $A \times A$, namely

$$\{(x, x) \; ; \; x \in A\}.$$

Given a mapping $f : A \to B$, consider the question : when does there exist a mapping $g : B \to A$ such that $g \circ f = \mathrm{id}_A$?

$$A \xrightarrow{\quad f \quad} B \xrightarrow{\quad g \quad} A$$
$$\underset{id}{\underline{\hspace{4cm}}}$$

Observe that such a mapping g does not exist in general.

Example Let $f : \mathbb{N} \to \mathbb{N}$ be given by

$$f(0) = f(1) = 1, \quad (\forall n \geq 2) \; f(n) = n.$$

There is no mapping $g : \mathbb{N} \to \mathbb{N}$ such that $g \circ f$ is the identity map on $\mathbb{N}$, for if such a map existed we would have the contradiction

$$0 = g[f(0)] = g(1),$$
$$1 = g[f(1)] = g(1).$$

2.3 Theorem *Concerning $f : A \to B$ the following statements are equivalent :*

(1) *there exists $g : B \to A$ such that $g \circ f = \mathrm{id}_A$;*

(2) $(\forall x, y \in A) \quad f(x) = f(y) \Longrightarrow x = y.$

Proof $(1) \Rightarrow (2)$: Suppose that (1) holds and let $x, y \in A$ be such that $f(x) = f(y)$. Then $g[f(x)] = g[f(y)]$ gives $x = y$.

$(2) \Rightarrow (1)$: Suppose now that (2) holds. Consider the set

$$G = \{(f(x), x) \; ; \; x \in A\} \subseteq \mathrm{Im}\, f \times A.$$

Observe that G is the graph of a mapping $t : \mathrm{Im}\, f \to A$. This follows from the fact that if $(y, z) \in G$ and $(y, z') \in G$ then there exist $x, x' \in A$ such that

$$y = f(x), z = x \text{ and } y = f(x'), z' = x'.$$

These give $f(x) = f(x')$ whence $x = x'$ by (2) and then $z = z'$. By definition, therefore, this mapping $t : \operatorname{Im} f \to A$ is then such that

$$(\forall x \in A) \quad x = t[f(x)].$$

Note, however, that t is not a mapping from B to A. We can remedy this deficiency by 'extending' t from $\operatorname{Im} f$ up to B as follows. Define $g : B \to A$ by

$$g(y) = \begin{cases} t(y) & \text{if } y \in \operatorname{Im} f; \\ \text{any } \alpha \in A & \text{otherwise.} \end{cases}$$

We then have

$$(\forall x \in A) \quad g[f(x)] = t[f(x)] = x,$$

i.e. $g \circ f = \operatorname{id}_A$. $\Diamond$

Definition A mapping $f : A \to B$ that satisfies either of the equivalent conditions of 2.3 is said to be *injective*. Any mapping $g : B \to A$ such that $g \circ f = \operatorname{id}_A$ is called a *left inverse* of f.

Example $f : \mathbb{N} \to \mathbb{N}$ given by $f(n) = n + 1$ is injective; for if $f(n) = f(m)$ then $n + 1 = m = 1$ gives $n = m$. There are infinitely many left inverses of f. In fact, for every $p \in \mathbb{N}$ define $g_p : \mathbb{N} \to \mathbb{N}$ by

$$g_p(n) = \begin{cases} n - 1 & \text{if } n \geq 1; \\ p & \text{if } n = 0. \end{cases}$$

Then for every $n \in \mathbb{N}$ we have

$$g_p[f(n)] = g_p(n + 1) = n,$$

so each g_p is a left inverse of f.

Example $f : \mathbb{N} \to \mathbb{N}$ given by $f(n) = n^2$ is injective, for $n^2 = m^2$ with $n, m \geq 0$ gives $n = m$.

Example $f : \mathbb{Z} \to \mathbb{N}$ given by $f(n) = n^2$ is not injective. For example, we have $f(1) = f(-1)$.

Example $f : \mathbb{R} \to \mathbb{R} \times \mathbb{R}$ given by $f(x) = (x, 0)$ is injective.

Example $f : E \to \mathbf{P}(E)$ given by $f(x) = \{x\}$ is injective.

Example $f : \mathbb{C} \to \mathbb{C}$ given by $f(x + iy) = x - iy$ is injective.

Example A function $f : \mathbb{R} \to \mathbb{R}$ is injective if and only if every line parallel to the x-axis meets the graph of f at most once.

Consider now the corresponding question. Given $f : A \to B$, when does there exist $g : B \to A$ such that $f \circ g = \mathrm{id}_B$?

$$B \xrightarrow{\quad g \quad} A \xrightarrow{\quad f \quad} B$$
$$\underset{id}{\underline{\qquad\qquad\qquad\qquad}}$$

Note that this situation is 'dual' to the previous one, in the sense that by reversing the arrows in the first we obtain the second.

Again observe that such a mapping g does not exist in general.

Example Let $f : \mathbb{R} \to \mathbb{R}$ be given by $f(x) = \sin x$. Then there is no mapping $g : \mathbb{R} \to \mathbb{R}$ such that $f \circ g$ is the identity map. For, if such a map g existed we would have, for example, $2 = f[g(2)] = \sin g(2)$ and this is impossible since $|\sin x| \leq 1$ for every x.

2.4 Theorem *Concerning $f : A \to B$ the following statements are equivalent :*

(1) *there exists $g : B \to A$ such that $f \circ g = \mathrm{id}_B$;*
(2) $\mathrm{Im}\, f = B.$

Proof $(1) \Rightarrow (2)$: If (1) holds then for every $x \in B$ we have

$$x = \mathrm{id}_B(x) = (f \circ g)(x) = f[g(x)] \in \mathrm{Im}\, f,$$

whence $B \subseteq \mathrm{Im}\, f$ and so $B = \mathrm{Im}\, f$.

$(2) \Rightarrow (1)$: If (2) holds then for every $y \in B$ there exists $x \in A$ such that $f(x) = y$. For each $y \in B$, choose once and for all an element x_y of A such that $f(x_y) = y$. Now define $g : B \to A$ by the prescription

$$(\forall y \in B) \quad g(y) = x_y.$$

Then for every $y \in B$ we have

$$f[g(y)] = f(x_y) = y,$$

so that $f \circ g = \mathrm{id}_B$. $\diamond$

Definition A mapping $f : A \to B$ that satisfies either of the equivalent conditions of 2.4 is said to be *surjective*. Any mapping $g : B \to A$ such that $f \circ g = \text{id}_B$ is a *right inverse* of f.

Example $f : \mathbb{N} \to \mathbb{N}$ given by

$$f(n) = \begin{cases} \frac{1}{2}n & \text{if } n \text{ is even;} \\ \frac{1}{2}(n-1) & \text{if } n \text{ is odd,} \end{cases}$$

is surjective. For, given any $m \in \mathbb{N}$ we have $f(2m) = m$. Note that f is not injective since we also have $f(2m+1) = m$. There are infinitely many right inverses of f. In fact, for every $p \in \mathbb{N}$ define $g_p : \mathbb{N} \to \mathbb{N}$ by

$$g_p(n) = \begin{cases} 2n+1 & \text{if } n \neq p; \\ 2p & \text{if } n = p. \end{cases}$$

Then we have

$$f[g_p(n)] = \begin{cases} f(2n+1) & \text{if } n \neq p; \\ f(2p) & \text{if } n = p, \end{cases}$$
$$= n \quad \text{in each case,}$$

so each g_p is a right inverse of f.

Example $f : \mathbb{Z} \to \mathbb{Z}$ given by $f(n) = n+1$ is surjective; for every $m \in \mathbb{Z}$ we have $f(m-1) = m$.

Example $f : \mathbb{N} \to \mathbb{N}$ given by $f(n) = n+1$ is not surjective; for there is no $n \in \mathbb{N}$ with $f(n) = 0$.

Example $f : \mathbb{R} \times \mathbb{R} \to \mathbb{R}$ given by $f(x,y) = x$ is surjective.

Example $f : \mathbb{C} \to \mathbb{C}$ given by $f(x+iy) = x - iy$ is surjective; for every $a + ib \in \mathbb{C}$ we have $f(a - ib) = a + ib$.

Example A function $f : \mathbb{R} \to \mathbb{R}$ is surjective if and only if every line parallel to the x-axis meets the graph of f at least once.

Combining the above notions, we obtain the following.

Definition A mapping $f : A \to B$ is called a *bijection* if it is both injective and surjective.

If $f : A \to B$ is a bijection then by 2.3 it has a left inverse $g : B \to A$, and by 2.4 it has a right inverse $h : B \to A$. Observe now that *in this situation g and h coincide*. To see this, note that $g \circ f = \mathrm{id}_A$ and $f \circ h = \mathrm{id}_B$ give, by the associativity of composition,

$$h = \mathrm{id}_A \circ h = (g \circ f) \circ h = g \circ (f \circ h) = g \circ \mathrm{id}_B = g.$$

Thus we can assert that if $f : A \to B$ is a bijection then there is a mapping $g : B \to A$ (necessarily also a bijection) such that $g \circ f = \mathrm{id}_A$ and $f \circ g = \mathrm{id}_B$. We can in fact say more than this : such a mapping g is *unique*. To see this, suppose that $\vartheta : B \to A$ is also such that $\vartheta \circ f = \mathrm{id}_A$ and $f \circ \vartheta = \mathrm{id}_B$. Then, again by the associativity of composition, we have

$$\vartheta = \mathrm{id}_A \circ \vartheta = (g \circ f) \circ \vartheta = g \circ (f \circ \vartheta) = g \circ \mathrm{id}_B = g.$$

This unique bijection g is called the *inverse* of the bijection f and henceforth will be denoted by f^{-1}.

Example $f : \mathbb{C} \to \mathbb{C}$ given by $f(x + iy) = x - iy$ is a bijection, for we have seen above that it is both injective and surjective. The inverse of f is f itself, for

$$f[f(x + iy)] = f(x - iy) = x + iy$$

shows that $f \circ f$ is the identity and so $f^{-1} = f$.

Example If $f : \mathbb{R} \to \mathbb{R}$ is a bijection then we have

$$(x, y) \in \text{graph of } f^{-1} \iff y = f^{-1}(x)$$
$$\iff f(y) = x$$
$$\iff (y, x) \in \text{graph of } f.$$

Thus the graph of f^{-1} is the reflection in the line $y = x$ of the graph of f.

Example It is shown in analysis that if $A = \{x \in \mathbb{R} \; ; \; x > 0\}$ then the *exponential function*

$$x \mapsto e^x = 1 + x + \frac{x^2}{2!} + \frac{x^3}{3!} + \cdots$$

is a bijection from $\mathbb{R}$ onto A. Its inverse is the *logarithmic function* from A to $\mathbb{R}$ given by $x \mapsto \log_e x$. The graph of each of these functions is the reflection in the line $y = x$ of the graph of the other.

More examples of injections, surjections, and bijections can be constructed using composite mappings, as the following result shows.

2.5 Theorem *Let* $A \xrightarrow{\;f\;} B \xrightarrow{\;g\;} C$ *be given mappings. Then*

(1) *if* f, g *are injective so also is* $g \circ f$;
(2) *if* f, g *are surjective so also is* $g \circ f$;
(3) *if* f, g *are bijective so also is* $g \circ f$, *and in this case*

$$(g \circ f)^{-1} = f^{-1} \circ g^{-1}.$$

Proof (1) If f, g are injective then clearly

$$g[f(x)] = g[f(y)] \Longrightarrow f(x) = f(y) \Longrightarrow x = y,$$

and so $g \circ f$ is injective.

(2) If f, g are surjective then for every $z \in C$ there exists $y \in B$ such that $g(y) = z$, then an $x \in A$ such that $f(x) = y$. Thus for every $z \in C$ there exists $x \in A$ such that $z = g[f(x)]$, so $g \circ f$ is surjective.

(3) If f, g are bijective then by (1) and (2) so is $g \circ f$. Moreover, since

$$g \circ f \circ f^{-1} \circ g^{-1} = g \circ \mathrm{id}_B \circ g^{-1} = g \circ g^{-1} = \mathrm{id}_C;$$
$$f^{-1} \circ g^{-1} \circ g \circ f = f^{-1} \circ \mathrm{id}_B \circ f = f^{-1} \circ f = \mathrm{id}_A,$$

we see that $(g \circ f)^{-1}$ is given by $f^{-1} \circ g^{-1}$. $\Diamond$

Definition Let $f : A \to B$ be a given mapping. If X is a subset of A then we shall denote by $f|_X : X \to B$ the mapping given by $f|_X(x) = f(x)$ for every $x \in X$, and call this the *restriction of f to the subset X*. Also, we shall denote by $f^+ : A \to \mathrm{Im}\, f$ the mapping given by $f^+(x) = f(x)$ for every $x \in A$.

Note that if $X \neq A$ then $f|_X \neq f$ since these mappings have different departure sets; and that f^+ is surjective by definition whereas f need not be.

2.6 Theorem *With every injection there is associated a bijection that has the same departure set.*

Proof If $f : A \to B$ is injective then by 2.3 there is a mapping $g : B \to A$ such that $g \circ f = \mathrm{id}_A$. The mapping $f^+ : A \to \mathrm{Im}\, f$ is then a bijection whose inverse is $g|_{\mathrm{Im}\, f}$. In fact, if $g^* = g|_{\mathrm{Im}\, f}$ then we have

$$(\forall x \in A) \qquad g^*[f^+(x)] = g^*[f(x)] = g[f(x)] = x,$$

so that $g^* \circ f^+ = \mathrm{id}_A$; and if $y = f(x) \in \mathrm{Im}\, f$ then $g(y) = g[f(x)] = x$ so

$$f^+[g^*(y)] = f^+[g(y)] = f[g(y)] = f(x) = y,$$

and hence $f^+ \circ g^* = \mathrm{id}_{\mathrm{Im}\, f}$. Thus $(f^+)^{-1} = g^*$. $\diamond$

Example $f : \mathbb{N} \to \mathbb{N}$ given by $f(n) = n + 1$ is injective. We have seen above that for every $p \in \mathbb{N}$ the mapping $g_p : \mathbb{N} \to \mathbb{N}$ given by

$$g_p(n) = \begin{cases} n - 1 & \text{if } n \geq 1; \\ p & \text{if } n = 0, \end{cases}$$

is a left inverse of f. Now $\mathrm{Im}\, f = \mathbb{N} \setminus \{0\}$, and $g_p^* = g_p|_{\mathrm{Im}\, f}$ is given by

$$g_p^*(n) = n - 1.$$

Clearly, g_p^* is the inverse of the bijection f^+.

2.7 Theorem *With every surjection there is associated a bijection that has the same arrival set.*

Proof If $f : A \to B$ is surjective then by 2.4 there is a mapping $g : B \to A$ such that $f \circ g = \mathrm{id}_B$. The mapping $f^* = f|_{\mathrm{Im}\, g} :$ $\mathrm{Im}\, g \to B$ is then a bijection whose inverse is $g^+ : B \to \mathrm{Im}\, g$. In fact,

$$(\forall y \in B) \qquad f^*[g^+(y)] = f^*[g(y)] = f[g(y)] = y,$$

so that $f^* \circ g^+ = \mathrm{id}_B$; and if $x = g(y) = g^+(y) \in \mathrm{Im}\, g^+$ then $f(x) = f[g(y)] = y$ so

$$g^+[f^*(x)] = g^+[f(x)] = g^+(y) = x,$$

and hence $g^+ \circ f^* = \mathrm{id}_{\mathrm{Im}\, g}$. Thus $(f^*)^{-1} = g^+$. $\diamond$

Example We have seen above that $f : \mathbb{N} \to \mathbb{N}$ given by

$$f(n) = \begin{cases} \frac{1}{2}n & \text{if } n \text{ is even;} \\ \frac{1}{2}(n-1) & \text{if } n \text{ is odd,} \end{cases}$$

is surjective, and that for every $p \in \mathbb{N}$ the mapping $g_p : \mathbb{N} \to \mathbb{N}$ given by

$$g_p(n) = \begin{cases} 2n + 1 & \text{if } n \neq p; \\ 2p & \text{if } n = p, \end{cases}$$

is a right inverse of f. If $\mathbb{N}^0$ denotes the set of odd natural numbers then $\mathrm{Im}\, g_p = \mathbb{N}^0 \cup \{2p\}$ and $f^* = f|_{\mathrm{Im}\, g_p}$ is given by

$$f^*(n) = \begin{cases} p & \text{if } n = 2p; \\ \frac{1}{2}(n-1) & \text{otherwise.} \end{cases}$$

Clearly, g_p^+ is the inverse of the bijection f^*.

We shall now show how the notion of a mapping can be used to introduce a convenient labelling device. Let I and E be sets and let $f : I \to E$ be a mapping, described by $i \mapsto f(i)$. We shall often write x_i instead of $f(i)$ and write the mapping as $(x_i)_{i \in I}$, which we call a *family of elements of E indexed by the set I*. Note that there may be repetitions in a family, in the sense that it is possible to have $x_i = x_j$ when $i \neq j$.

The most common indexing set is the set $\mathbb{N}$ of natural numbers. A set of elements of E indexed by $\mathbb{N}$, i.e. a mapping $f : \mathbb{N} \to E$ described by $n \mapsto x_n$, is called a *sequence of elements of E* and is written $(x_n)_{n \in \mathbb{N}}$ or $(x_n)_{n \geq 0}$. Thus, to obtain a sequence of elements of E it is equivalent to choose elements of E and label them $x_0, x_1, x_2, \ldots$.

Example The *Fibonacci sequence* is given by $(x_i)_{i \geq 0}$ where

$$x_0 = 0, \ x_1 = 1 \text{ and } (\forall n \geq 1) \ x_{n+1} = x_n + x_{n-1},$$

i.e. it is the sequence $0, 1, 1, 2, 3, 5, 8, 13, 21, \ldots$.

If in the above definition we replace E by $\mathbf{P}(E)$ then the mapping $f : I \to \mathbf{P}(E)$ may be described by $i \mapsto A_i$ where each A_i is a subset of E. This gives a *family* $(A_i)_{i \in I}$ *of subsets of E.*

Definition If $(A_i)_{i \in I}$ is a family of subsets of a set E then the *intersection* of the family is the subset

$$\bigcap_{i \in I} A_i = \{x \in E \ ; \ (\forall i \in I) \ x \in A_i\},$$

i.e. it is the subset consisting of those elements of E that belong to every set in the family; and the *union* of the family is the subset

$$\bigcup_{i \in I} A_i = \{x \in E \ ; \ (\exists i \in I) \ x \in A_i\},$$

i.e. it is the subset consisting of those elements of E that belong to at least one set in the family.

Example If $I = \{1, 2\}$ then $\bigcap_{i \in I} A_i = A_1 \cap A_2$ and $\bigcup_{i \in I} A_i = A_1 \cup A_2$.

Example If, for every real number $r \geq 0$,

$$A_r = \{(x, y) \in \mathbb{R} \times \mathbb{R} \ ; \ x^2 + y^2 \leq r\}$$

then $\bigcap_{r \geq 0} A_r = \{(0,0)\}$ and $\bigcup_{r \geq 0} A_r = \mathbb{R} \times \mathbb{R}$.

Example If E is a non-empty set and $(A_x)_{x \in E}$ is given by $A_x = \{x\}$ for every $x \in E$ then $\bigcap_{x \in E} A_x = \emptyset$ and $\bigcup_{x \in E} A_x = E$.

Example For $(m, n) \in \mathbb{Z} \times \mathbb{Z}$ consider the square region

$$R_{(m,n)} = \{(x, y) \in \mathbb{R} \times \mathbb{R} \ ; \ m \leq x \leq m+1, \ n \leq y \leq n+1\}.$$

We have

$$\bigcup_{i \in \mathbb{Z}} R_{(m,i)} \cap \bigcup_{j \in \mathbb{Z}} R_{(j,n)} = R_{(m,n)}.$$

Also,

$$\bigcup_{n \in \mathbb{Z}} R_{(n,n)} \cap \bigcup_{n \in \mathbb{Z}} R_{(n,-n)} = R_{(0,0)}.$$

Intersections and unions of families of subsets satisfy properties that generalize the familiar properties of finite intersections and unions. For example, in the next result we establish the *general associativity* of intersection and of union. Roughly speaking, it says that in order to obtain the intersection (union) of a family of sets we can write the family with an arbitrary grouping of the terms and replace each group by its intersection (union).

2.8 Theorem *Let $(A_i)_{i \in I}$ be a family of sets and let $I = \bigcup_{j \in J} I_j$.*

Then

$$\bigcap_{i \in I} A_i = \bigcap_{j \in J} \left(\bigcap_{i \in I_j} A_i \right), \quad \bigcup_{i \in I} A_i = \bigcup_{j \in J} \left(\bigcup_{i \in I_j} A_i \right).$$

Proof For every $j \in J$ let $B_j = \bigcap_{i \in I_j} A_i$. Then

$$x \in \bigcap_{i \in I} A_i \iff (\forall i \in I)\, x \in A_i$$

$$\iff (\forall j \in J)(\forall i \in I_j)\, x \in A_i$$

$$\iff (\forall j \in J)\, x \in \bigcap_{i \in I_j} A_i = B_j$$

$$\iff x \in \bigcap_{j \in J} B_j,$$

from which the first equality follows. The second is established in a similar way. ◇

Distributivity can also be generalized.

2.9 Theorem *If $(A_i)_{i \in I}$ is a family of sets and X is any set then*

$$X \cap \bigcup_{i \in I} A_i = \bigcup_{i \in I} (X \cap A_i);$$

$$X \cup \bigcap_{i \in I} A_i = \bigcap_{i \in I} (X \cup A_i).$$

Proof We establish the first equality, the proof of the second being similar. Now $x \in X \cap \bigcup_{i \in I} A_i$ if and only if $x \in X$ and, for some $i \in I$, $x \in A_i$; and this is so if and only if, for some $i \in I$, $x \in X \cap A_i$. ◇

Finally, the de Morgan laws generalize as follows.

2.10 Theorem *If $(A_i)_{i \in I}$ is a family of subsets of a set E then*

$$\left(\bigcap_{i \in I} A_i \right)' = \bigcup_{i \in I} A_i', \quad \left(\bigcup_{i \in I} A_i \right)' = \bigcap_{i \in I} A_i'.$$

Proof The first equality follows from the fact that, by 2.9,

$$\bigcap_{i \in I} A_i \cap \bigcup_{j \in I} A_j' = \bigcup_{j \in I} \left(\bigcap_{i \in I} A_i \cap A_j' \right) = \bigcup_{j \in I} \emptyset = \emptyset;$$

$$\bigcap_{i \in I} A_i \cup \bigcup_{j \in I} A_j' = \bigcap_{i \in I} \left(A_i \cup \bigcup_{j \in I} A_j' \right) = \bigcap_{i \in I} E = E,$$

and the second is established similarly. $\Diamond$

We continue our consideration of mappings by concentrating on the important notion of a bijection.

Definition Sets E and F are said to be *equipotent* if there is a bijection $f : E \to F$.

Example $E \times F$ and $F \times E$ are equipotent. In fact, the mapping $f : E \times F \to F \times E$ given by the assignment

$$(x, y) \mapsto (y, x)$$

is clearly a bijection.

Example If $f : A \to B$ is injective then A and $\operatorname{Im} f$ are equipotent. In fact, as shown in the proof of 2.6, the mapping $f^+ : A \to \operatorname{Im} f$ is a bijection.

Example For every set A the power set $\mathbf{P}(A)$ is equipotent to the set of mappings from A to the two-element set $\{0, 1\}$. To see this, first define for every subset B of A the *characteristic function* $\chi_B : A \to \{0, 1\}$ by

$$\chi_B(x) = \begin{cases} 1 & \text{if } x \in B; \\ 0 & \text{if } x \notin B. \end{cases}$$

Then, if 2^A denotes the set of all mappings from A to the two-element set $\{0, 1\}$, we have $\chi_B \in 2^A$ and we can define a mapping $\vartheta : \mathbf{P}(A) \to 2^A$ by the prescription

$$\vartheta(B) = \chi_B.$$

We claim that ϑ is a bijection. To show this, it suffices to find an inverse for ϑ. For this purpose, given any $f \in 2^A$ let

$$f^{\leftarrow}\{1\} = \{x \in A \; ; \; f(x) = 1\}.$$

Then we can define a mapping $\mu : 2^A \to \mathbf{P}(A)$ by the prescription

$$\mu(f) = f^{\leftarrow}\{1\}.$$

We claim that $\mu = \vartheta^{-1}$. To prove this, observe that for every $x \in A$ we have

$$\chi_{f^{\leftarrow}\{1\}}(x) = \begin{cases} 1 & \text{if } x \in f^{\leftarrow}\{1\}; \\ 0 & \text{if } x \notin f^{\leftarrow}\{1\}, \end{cases}$$

$$= \begin{cases} 1 & \text{if } f(x) = 1; \\ 0 & \text{if } f(x) = 0, \end{cases}$$

$$= f(x),$$

and therefore, for every $f \in 2^A$,

$$\chi_{f^{\leftarrow}\{1\}} = f.$$

It follows that

$$(\vartheta \circ \mu)(f) = \vartheta[f^{\leftarrow}\{1\}] = \chi_{f^{\leftarrow}\{1\}} = f,$$

and so $\vartheta \circ \mu$ is the identity mapping on 2^A. Also, it is clear that for every $B \in \mathbf{P}(A)$ we have

$$\chi_B^{\leftarrow}\{1\} = B,$$

and therefore

$$(\mu \circ \vartheta)(B) = \mu(\chi_B) = \chi_B^{\leftarrow}\{1\} = B,$$

so that $\mu \circ \vartheta$ is the identity mapping on $\mathbf{P}(A)$. We conclude that ϑ is a bijection with $\vartheta^{-1} = \mu$, so $\mathbf{P}(A)$ and 2^A are equipotent.

The important feature of a bijection is that it allows us to compare two sets. It is intuitively clear that if A, B are finite sets then they are equipotent if and only if they have the same number of elements. The notion of a bijection can also be used to compare infinite sets, though we must postpone a discussion of this until later when we shall be able to tackle properly the question of what is meant by an 'infinite' set and how to compare the 'sizes' of two such sets. An important step in this direction, however, is the following result, in the proof of which we make use of some special notation.

To every mapping $f : A \to B$ we can associate a mapping $f^\to : \mathbf{P}(A) \to \mathbf{P}(B)$ by the prescription

$$f^\to(X) = \{f(x) \; ; \; x \in X\};$$

and a mapping $f^\leftarrow : \mathbf{P}(B) \to \mathbf{P}(A)$ by the prescription

$$f^\leftarrow(Y) = \{x \in A \; ; \; f(x) \in Y\}.$$

Definition We call $f^\to$ the *direct image map* and $f^\leftarrow$ the *inverse image map* induced by f.

Observe that $f^\to$ and $f^\leftarrow$ are *inclusion preserving*, in the sense that

$$X_1 \subseteq X_2 \implies f^\to(X_1) \subseteq f^\to(X_2);$$
$$Y_1 \subseteq Y_2 \implies f^\leftarrow(Y_1) \subseteq f^\leftarrow(Y_2).$$

These properties follow immediately from the definitions.

2.11 Theorem [Schröder-Bernstein] *If there exist injections* $f : A \to B$ *and* $g : B \to A$ *then* A *and* B *are equipotent.*

Proof Denoting the operation of taking complements in A by C_A, we have the mapping $C_A : \mathbf{P}(A) \to \mathbf{P}(A)$ which is inclusion inverting in the sense that

$$X_1 \subseteq X_2 \implies C_A(X_1) \supseteq C_A(X_2).$$

Similarly, so is the mapping $C_B : \mathbf{P}(B) \to \mathbf{P}(B)$.

Consider the composite mapping

$$\varsigma = C_A \circ g^\to \circ C_B \circ f^\to.$$

Clearly, we have $\varsigma : \mathbf{P}(A) \to \mathbf{P}(A)$. Since $f^\to, g^\to$ are inclusion preserving and C_A, C_B are inclusion inverting, it follows that ς is inclusion preserving.

Consider now the set

$$F = \{X \in \mathbf{P}(A) \; ; \; X \subseteq \varsigma(X)\}.$$

Note that $F \neq \emptyset$ since it contains the empty subset of A. Let

$$G = \bigcup\{X \; ; \; X \in F\}$$

be the union of all the sets in F. Then for every $X \in F$ we have

$$X \subseteq \varsigma(X) \text{ and } X \subseteq G.$$

Since ς is inclusion preserving we then have

$$X \subseteq \varsigma(X) \subseteq \varsigma(G).$$

It follows that $\varsigma(G)$ contains the union of all the sets X in F, i.e. that

$$(\star) \hspace{4cm} G \subseteq \varsigma(G).$$

Again since ς is inclusion preserving, we then have

$$\varsigma(G) \subseteq \varsigma[\varsigma(G)],$$

and therefore $\varsigma(G) \in F$. Consequently, by the definition of G, we see that

$$(\star\star) \hspace{4cm} \varsigma(G) \subseteq G.$$

It follows from $(\star)$ and $(\star\star)$ that $G = \varsigma(G)$. Thus, by the definition of ς, we have that

$$C_A(G) = (C_A \circ \varsigma)(G) = (g^{\rightarrow} \circ C_B \circ f^{\rightarrow})(G),$$

which we can describe pictorially as follows :

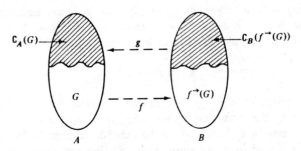

Now since, by hypothesis, f and g are injective this configuration shows that we can define a bijection $h : A \to B$ by the prescription

$$h(x) = \begin{cases} f(x) & \text{if } x \in G; \\ \text{the unique element of } g^{\leftarrow}\{x\} & \text{if } x \notin G. \end{cases}$$

Thus A and B are equipotent. $\Diamond$

Example Any two closed intervals of $\mathbb{R}$ are equipotent. To see this, let $a, b, c, d \in \mathbb{R}$ be such that $a < b$ and $c < d$. Consider the mapping

$$f : [a, b] \to [c, d]$$

given by the prescription

$$f(x) = \frac{(b - x)c + (x - a)d}{b - a}.$$

Note that if $a \le x \le b$ then $\lambda = \dfrac{b - x}{b - a} < 1$ and $\mu = \dfrac{x - a}{b - a} < 1$ with $\lambda + \mu = 1$, so $f(x) \in [c, d]$. If now $f(x) = f(y)$ then

$$(b - x)c + (x - a)d = (b - y)c + (y - a)d$$

gives $x(d - c) = y(d - c)$ whence $x = y$. Thus f is injective. Similarly, we can define an injection

$$g : [c, d] \to [a, b].$$

It follows by the Schröder-Bernstein Theorem that $[a, b]$ and $[c, d]$ are equipotent.

Example The intervals $[a, b],]a, b[, [a, b[,]a, b]$ are equipotent. To see this, let $c, d \in \mathbb{R}$ be such that $a < c < d < b$. As in the preceding Example, there is an injection $f : [a, b] \to [c, d]$. If

$$i : [c, d] \to]a, b[$$

is the restriction of $\mathrm{id}_{]a,b[}$ to $[c, d]$ then $i \circ f$ is an injection of $[a, b]$ into $]a, b[$. Since the restriction of $\mathrm{id}_{[a,b]}$ to $]a, b[$ is an injection of $]a, b[$ into $[a, b]$, it follows by the Schröder-Bernstein Theorem that $[a, b]$ and $]a, b[$ are equipotent. Similar arguments with $]a$ replaced by $[a,$ and $b[$ replaced by $b]$, show that $[a, b]$ is also equipotent to $[a, b[$ and to $]a, b]$.

Example Here we describe explicitly a bijection

$$f : [-1, 1] \to \,]-1, 1[\,.$$

Consider the prescription

$$f(x) = \begin{cases} \frac{1}{2}x & \text{if } x = \pm\frac{1}{2^n}; \\ x & \text{otherwise.} \end{cases}$$

It is readily verified that f is a bijection. Thus $[-1, 1]$ is equipotent to $]-1, 1[$.

Example The open interval $]-\frac{1}{2}\pi, \frac{1}{2}\pi[$ is equipotent to $\mathbb{R}$. In fact, the mapping

$$f : \,]-\tfrac{1}{2}\pi, \tfrac{1}{2}\pi[\,\to \mathbb{R}$$

given by $f(\vartheta) = \tan \vartheta$ is a bijection.

The above Examples show that for all $a, b \in \mathbb{R}$ with $a < b$ the interval $[a, b]$ is equipotent to $\mathbb{R}$.

CHAPTER THREE

Equivalence relations

We now consider the simple, but very important, notion of a collection of pairwise disjoint subsets that fit together like a jig-saw puzzle.

Definition By a *partitioning* of a set E we shall mean a family $(A_i)_{i \in I}$ of non-empty subsets of E such that

(1) $\bigcup_{i \in I} A_i = E$;

(2) $(i \neq j) \ A_i \cap A_j = \emptyset$.

Note that if $(A_i)_{i \in I}$ is a partitioning of E then the subsets A_i are distinct, for if we had $A_i = A_j$ for $i \neq j$ then (2) would give the contradiction $A_i = \emptyset$.

Definition If $(A_i)_{i \in I}$ is a partitioning of E then we shall say that the set $\{A_i \ ; \ i \in I\}$ of associated subsets is a *partition* of E.

Note that a partitioning is by definition a mapping $f : I \rightarrow \mathbf{P}(E)$. A partition is thus the 'effect' of this mapping, namely a selection of non-empty pairwise disjoint subsets of E whose union is the whole of E. The analogy with a jig-saw puzzle is useful to bear in mind; the subsets A_i are the pieces of the puzzle.

Example If A is a subset of E with $A \neq \emptyset$ and $A \neq E$ then $\{A, A'\}$ is a partition of E.

Example If $E = \{a, b, c, d\}$ then $\{\{a\}, \{b, c\}, \{d\}\}$ is a partition of E. So also is $\{\{a, b\}, \{c, d\}\}$.

Example If A is the set of even integers and B is the set of odd integers then $\{A, B\}$ is a partition of $\mathbb{Z}$.

Example If $E_m^* = \{(x, mx) \; ; \; x \in \mathbb{R}, x \neq 0\}$ is the line of gradient m that is 'punctured' at the origin then $\{E_m^* \; ; \; m \in \mathbb{R}\}$ together with the y-axis is a partition of $\mathbb{R} \times \mathbb{R}$.

Example If C_r denotes the circle of radius r centred at $(0, 0)$ then $\{C_r \; ; \; r \geq 0\}$ is a partition of $\mathbb{R} \times \mathbb{R}$.

The notion of a partition is closely related to a particular type of relation which is a very important tool in mathematics.

Definition By an *equivalence relation* on a set E we shall mean a relation R on E which is

(1) *reflexive*, in the sense that xRx for every $x \in E$;

(2) *symmetric*, in the sense that if xRy then yRx;

(3) *transitive*, in the sense that if xRy and yRz then xRz.

Example Let E be the set of lines in $\mathbb{R} \times \mathbb{R}$. Define a relation R on E by

$$x \, R \, y \iff x \text{ is parallel to } y.$$

Then R is an equivalence relation on E.

Example Given any mapping $f : E \to F$, let R_f be the relation defined on E by

$$x \, R_f \, y \iff f(x) = f(y).$$

Then R_f is an equivalence relation on E. We call it the *equivalence relation associated with the mapping* f.

Example Let 'mod n' be the relation $\equiv_n$ defined on $\mathbb{Z}$ by

$$x \equiv_n y \iff (\exists k \in \mathbb{Z}) \; x - y = kn,$$

i.e. x and y are equivalent modulo n if and only if they differ by an integer multiple of n. Then this is an equivalence relation on $\mathbb{Z}$. In fact,

(1) for every $x \in \mathbb{Z}$ we have $x - x = 0 = 0n$, so $x \equiv_n x$;

(2) if $x \equiv_n y$ then $x - y = kn$ for some $k \in \mathbb{Z}$, so $y - x = (-k)n$ and therefore $y \equiv_n x$;

(3) if $x \equiv_n y$ and $y \equiv_n z$ then $x - y = kn$ and $y - z = tn$ for some $k, t \in \mathbb{Z}$, so $x - z = (x - y) + (y - z) = (k + t)n$ and therefore $x \equiv_n z$.

Example Let S be a set of sets. Recall that $A, B \in S$ are said to be equipotent if there is a bijection $f : A \to B$. The relation of being equipotent is an equivalence relation on S. The transitivity of this relation is a consequence of 2.5(3).

Example The relation $\sim$ defined on $\mathbb{R}$ by

$$x \sim y \iff x - y \in \mathbb{Q}$$

is an equivalence relation.

Example The relation $\sim$ defined on $\mathbb{R} \times \mathbb{R}$ by

$$(a, b) \sim (c, d) \iff a^2 + b^2 = c^2 + d^2$$

is an equivalence relation.

If R is an equivalence relation on a set E then for every $x \in E$ we shall write

$$[x]_R = \{y \in E \; ; \; yRx\}$$

and call this the R-*class* of the element x. Other common notation is x/R and R_x.

3.1 Theorem *If R is an equivalence relation on E then the following statements are equivalent :*

(1) yRx;
(2) $y \in [x]_R$;
(3) $[y]_R = [x]_R$.

Proof (1) $\iff$ (2) : This is immediate from the definition of the R-class $[x]_R$.

(1) $\Rightarrow$ (3) : Suppose that (1) holds and let $y^* \in [y]_R$. Then yRx and y^*Ry. By the transitivity of R, we deduce that y^*Rx, i.e. that $y^* \in [x]_R$. Thus we see that $[y]_R \subseteq [x]_R$. Using the symmetry of R, we can interchange x and y in this argument and deduce also that $[x]_R \subseteq [y]_R$, whence we have (3).

(3) $\Rightarrow$ (2) : The reflexivity of R gives yRy, so that $y \in [y]_R$. Thus, if (3) holds then $y \in [y]_R = [x]_R$. $\Diamond$

3.2 Theorem *If R is a equivalence relation on E then any two distinct R-classes are disjoint.*

Proof Suppose that $[x]_R$ and $[y]_R$ are not disjoint. Then they have an element in common, say z. From $z \in [x]_R$ we obtain, by 3.1, $[z]_R = [x]_R$; and from $z \in [y]_R$ we obtain $[z]_R = [y]_R$. Thus we have $[x]_R = [y]_R$. It follows, therefore, that if $[x]_R \neq [y]_R$ then we must have $[x]_R \cap [y]_R = \emptyset$. $\diamond$

The connection between partitions and equivalence relations is as follows.

3.3 Theorem *If R is an equivalence relation on a set E then the collection of R-classes is a partition of E. Conversely, every partition of E determines an equivalence relation on E.*

Proof Suppose that R is an equivalence relation on E. By the reflexivity of R and 3.1, we have $x \in [x]_R$ for every $x \in E$. It follows that E is the union of all the R-classes. Now, by 3.2, the R-classes are pairwise disjoint. Hence the R-classes form a partition of E.

Conversely, suppose that a collection P of subsets of E is a partition of E. Define a relation R on E by

$$x\,R\,y \iff \begin{cases} x, y \text{ belong to the same 'piece'} \\ \text{in the jig-saw puzzle } P. \end{cases}$$

Then it is readily seen that R is an equivalence relation on E, the R-classes being precisely the 'pieces' in the jig-saw puzzle P. $\diamond$

Example There are five distinct equivalence relations definable on the set $E = \{1, 2, 3\}$. In fact, each equivalence relation defines a partition and there are only five partitions possible, namely the 1-piece jigsaw puzzle $\{E\}$, the three 2-piece jig-saw puzzles

$$\{\{1\}, \{2, 3\}\}, \quad \{\{1, 2\}, \{3\}\}, \quad \{\{1, 3\}, \{2\}\},$$

and the 3-piece jig-saw puzzle $\{\{1\}, \{2\}, \{3\}\}$.

Example The partition associated with the equivalence relation 'mod n' has n classes, namely

$$[0]_n = \{\ldots, -2n, -n, 0, n, 2n, \ldots\},$$
$$[1]_n = \{\ldots, -2n + 1, -n + 1, 1, n + 1, 2n + 1, \ldots\},$$
$$\vdots$$
$$[n - 1]_n = \{\ldots, -n - 1, -1, n - 1, 2n - 1, 3n - 1, \ldots\}.$$

Example The relation $\sim$ defined on $\mathbb{Z}$ by

$$m \sim n \iff |m| = |n|$$

ia an equivalence relation. The $\sim$-class of n is $\{n, -n\}$.

Example For every $x \in \mathbb{R}$ let $[\![x]\!]$ denote the greatest integer y such that $y \le x$. Then the relation $\sim$ defined on $\mathbb{R}$ by

$$x \sim y \iff [\![x]\!] = [\![y]\!]$$

is an equivalence relation. The $\sim$-class of $x \in \mathbb{R}$ is the interval $[n, n+1[$ where $n = [\![x]\!]$. The corresponding partition of $\mathbb{R}$ is

$$\{[n, n+1[\; ; \; n \in \mathbb{Z}\}.$$

Example Let $\sim$ be the relation defined on $\mathbb{R} \times \mathbb{R}$ by

$$(x, y) \sim (a, b) \iff y - |x| = b - |a|.$$

Then $\sim$ is an equivalence relation. Given any $(a, b) \in \mathbb{R} \times \mathbb{R}$, let $b - |a| = t$. Then the $\sim$-class of (a, b) is

$$A_t = \{(x, y) \in \mathbb{R} \times \mathbb{R} \; ; \; y = |x| + t\},$$

which is the half-line pair

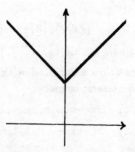

the apex being at the point $(0, t)$. The corresponding partition is the set $\{A_t \; ; \; t \in \mathbb{R}\}$.

Example Let $E = \{(x, y) \in \mathbb{R} \times \mathbb{R} \; ; \; x \neq 0, y \neq 0\}$ be the cartesian plane with the axes deleted. Consider the relation $\equiv$ defined on E by

$$(x, y) \equiv (a, b) \iff xy(a^2 + b^2) = ab(x^2 + y^2).$$

Because E does not contain $(0, 0)$, we can rewrite this in the form

$$(x, y) \equiv (a, b) \iff \frac{xy}{x^2 + y^2} = \frac{ab}{a^2 + b^2},$$

from which it is clear that $\equiv$ is an equivalence relation.

To describe the $\equiv$-classes, consider first an element $(1, m)$ where $m \neq 0$. We have

$$
\begin{aligned}
(x, y) \equiv (1, m) &\iff \frac{xy}{x^2 + y^2} = \frac{m}{1 + m^2} \\
&\iff xy + xym^2 = mx^2 + my^2 \\
&\iff (mx - y)(x - my) = 0 \\
&\iff y = mx \text{ or } y = \frac{x}{m}.
\end{aligned}
$$

Thus the $\equiv$-class of $(1, m)$ is the pair of lines $y = mx$ and $y = x/m$ with, of course, the origin deleted.

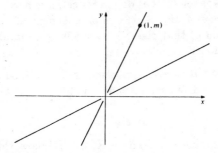

As for the $\equiv$-class of a general point $(a, b) \in E$, observe from the definition of $\equiv$ that we have

$$(a, b) \equiv \left(1, \frac{b}{a}\right),$$

so the $\equiv$-class of (a, b) is the $\equiv$-class of $\left(1, \frac{b}{a}\right)$.

Definition If R is an equivalence relation on E then the set of R-classes is often called the *quotient set* of E by R and is denoted by E/R.

The quotient set is thus just another name for the associated partition, and so we have

$$E/R = \{[x]_R \; ; \; x \in E\}.$$

Because the elements of E/R fail to have a unique representation, in the sense that if $y \in [x]_R$ then $[y]_R = [x]_R$, care must be taken when defining a mapping with E/R as departure set.

Example Let R be the equivalence relation 'mod 4' on $\mathbb{Z}$. Consider the prescription

$$\vartheta([x]_R) = x.$$

This does not define a mapping from $\mathbb{Z}/R$ to $\mathbb{Z}$ since, for example,

$$[1]_R = [5]_R$$

whereas

$$\vartheta([1]_R) = 1 \neq 5 = \vartheta([5]_R).$$

In order to check that a given prescription defines a mapping with a quotient set E/R as departure set, it is necessary to show that the image of any R-class is independent of the particular representation of that class. The following example illustrates how this is done.

Example Let R be the equivalence relation 'mod 4' on $\mathbb{Z}$ and let S be the equivalence relation 'mod 2' on $\mathbb{Z}$. Then the prescription

$$\vartheta([x]_R) = [x]_S$$

defines a mapping $\vartheta : \mathbb{Z}/R \to \mathbb{Z}/S$. To show this, take two representations $[x]_R$ and $[x']_R$ of the same R-class. Then $[x]_R = [x']_R$ so $x \equiv_R x'$ and therefore

$$x = x' + 4\lambda$$

for some $\lambda \in \mathbb{Z}$. Then

$$x = x' + 2\mu$$

where $\mu = 2\lambda \in \mathbb{Z}$ and so $x \equiv_S x'$ whence $[x]_S = [x']_S$. Thus the images under ϑ are independent of the representation of classes in the departure set $\mathbb{Z}/R$ and so ϑ does define a mapping from $\mathbb{Z}/R$ to $\mathbb{Z}/S$.

We can define in a natural way a mapping from E to E/R by the assignment $x \mapsto [x]_R$. Clearly, this mapping is surjective. We call it the *natural* (or *canonical*) *surjection* from E onto E/R and denote it by

$$\natural_R : E \to E/R.$$

Consider now a mapping $f : E \to F$ and the associated equivalence relation R_f. Recall that R_f is given by

$$x \equiv y(R_f) \iff f(x) = f(y).$$

The R_f-classes consist of elements that have the same image under f, and are singletons if and only if f is injective. The quotient set E/R_f plays an important rôle in the proof of the following result.

3.4 Theorem *Every mapping can be expressed as a composite of an injection, a bijection, and a surjection.*

Proof Given $f : E \to F$, consider the sets E/R_f and $\operatorname{Im} f$. Observing that

$$(\star) \qquad [x]_{R_f} = [y]_{R_f} \implies x \equiv y(R_f) \implies f(x) = f(y),$$

we can define a mapping

$$\vartheta : E/R_f \to \operatorname{Im} f$$

by the prescription

$$\vartheta([x]_{R_f}) = f(x).$$

Clearly, ϑ is surjective. Observing that the converse implications in $(\star)$ also hold, namely that

$$f(x) = f(y) \implies x \equiv y(R_f) \implies [x]_{R_f} = [y]_{R_f},$$

we deduce from 2.3 that ϑ is also injective. Thus ϑ is a bijection.

Now let $\natural : E \to E/R_f$ be the natural surjection and let $i : \operatorname{Im} f \to F$ be the restriction of id_F to $\operatorname{Im} f$. The composite mapping $i \circ \vartheta \circ \natural : E \to F$, namely

$$E \xrightarrow{\;\natural\;} E/R_f \xrightarrow{\;\vartheta\;} \operatorname{Im} f \xrightarrow{\;i\;} F,$$

can be described by

$$x \mapsto [x]_{R_f} \mapsto f(x) \mapsto f(x).$$

Clearly, this is the same as f, so we have the decomposition

$$f = i \circ \vartheta \circ \natural,$$

which expresses f as a composite of an injection, a bijection, and a surjection. ◇

It should be noted that the above decomposition is not unique. The particular decomposition described in the above proof is called the *canonical decomposition* of f.

3.5 Corollary *Every mapping can be expressed as a composite of an injection and a surjection.*

Proof Every bijection is both an injection and a surjection, so the result follows by 2.5(1) or 2.5(2). In fact, observe that with the above notation we have $\vartheta \circ \natural = f^+$ so $f = i \circ f^+$. Also, if $b = i \circ \vartheta$ then b is injective and $f = b \circ \natural$. ◇

The integers

We now turn our attention to a specific set, namely the set $\mathbb{Z}$ of integers. The subset $\mathbb{N}$ of natural numbers will play an important part in the discussion. We begin by considering the notion of an ordered set.

Definition A relation R on a set E is said to be *anti-symmetric* if the conditions xRy and yRx imply that $x = y$. By an *order* on E we mean a relation R on E that is reflexive, anti-symmetric, and transitive.

We shall generally denote an order relation by the symbol $\leq$. Thus, $\leq$ is an order on E if

(1) $(\forall x \in E) \ x \leq x$;
(2) if $x \leq y$ and $y \leq x$ then $x = y$;
(3) if $x \leq y$ and $y \leq z$ then $x \leq z$.

Definition An order $\leq$ on E is said to be a *total order*, and E is said to form a *chain*, if for all $x, y \in E$ either $x \leq y$ or $y \leq x$.

Example The relation $\subseteq$ of set inclusion is an order on $\mathbf{P}(E)$.

Example Consider the *divisibility relation* defined on $\mathbb{N}$ by

$$a \leq b \iff a \text{ divides } b.$$

It is clear that this relation is reflexive, anti-symmetric, and transitive on $\mathbb{N}$ and therefore is an order.

Example The set $\mathbb{N}$ of natural numbers can also be ordered by the relation
$$a \leq b \iff b - a \in \mathbb{N}.$$

This order is a total order. When $a \leq b$ and $a \neq b$ we shall write $a < b$ and say that a is *less than* b. We then have the familiar chain

$$0 < 1 < 2 < 3 < \cdots < n < n+1 < \cdots.$$

The total order on $\mathbb{N}$ described in the previous Example is often referred to as the *natural order* on $\mathbb{N}$. An important property of this order is the following, which we shall take as an axiom : $\mathbb{N}$ is *well ordered*, in the sense that *every non-empty subset of $\mathbb{N}$ has a least element*.

Note that, with respect to its obvious natural order, the set $\mathbb{Z}$ of integers is not well ordered in the above sense; for the subset of negative integers has no least element.

An important consequence of the well ordering axiom of $\mathbb{N}$ is the following result.

4.1 Theorem [Principle of induction] *Given* $m \in \mathbb{N}$, *let*

$$S_m = \{x \in \mathbb{N} \; ; \; x \geq m\}.$$

If S is a subset of S_m such that

(1) $m \in S$;
(2) $n \in S$ *implies* $n+1 \in S$,

then we have $S = S_m$.

Proof Since we are given that $S \subseteq S_m$, it suffices to prove that $S_m \setminus S = \emptyset$. Suppose, by way of obtaining a contradiction, that $A = S_m \setminus S$ is not empty. Then, by the well ordering axiom, A contains a least element, t say. Note that $t \neq m$ since $m \in S$ so that $m \notin A$. We therefore have that $t - 1 \geq m$ and so, by the minimality of t, $t - 1 \notin A$ whence $t - 1 \in S$. We now use property (2) to see that $t = (t - 1) + 1 \in S$ and hence that $t \notin A = S_m \setminus S$, a contradiction. We conclude, therefore, that $S_m \setminus S = \emptyset$ and thus $S = S_m$ as required. $\diamond$

The principle of induction is a very useful tool, as we shall now illustrate.

Example For every $n \in \mathbb{N}$ we have

$$(\star) \qquad \sum_{i=0}^{n} i = \tfrac{1}{2}n(n + 1).$$

Since we want to prove that this formula is true for all $n \in \mathbb{N}$, we take $m = 0$ in 4.1 so that $S_m = \mathbb{N}$. Let S be the subset of $\mathbb{N}$ consisting of those natural numbers n for which $(\star)$ holds. We show that $S = \mathbb{N}$ by induction, using (1) and (2) of 4.1.

(1) $\displaystyle\sum_{i=0}^{0} i = 0 = \frac{1}{2} \cdot 0 \cdot (0 + 1);$

(2) Suppose that $n \in S$, so that

$$\sum_{i=0}^{n} i = \frac{1}{2} n(n + 1).$$

We have to show that $n + 1 \in S$. Now

$$\begin{aligned}
\sum_{i=0}^{n+1} i &= \sum_{i=0}^{n} i + (n + 1) \\
&= \tfrac{1}{2} n(n + 1) + (n + 1) \\
&= (n + 1)(\tfrac{1}{2} n + 1) \\
&= \tfrac{1}{2}(n + 1)(n + 2).
\end{aligned}$$

Thus $n + 1 \in S$ and the proof is complete.

Example Let E_n be a set with $|E_n| = n$. We shall prove by induction that $|\mathbf{P}(E_n)| = 2^n$. For this purpose, let

$$S = \{n \in \mathbb{N} \; ; \; |\mathbf{P}(E_n)| = 2^n\}.$$

We show that $S = \mathbb{N}$. First note that $0 \in S$; for if we take $E_0 = \emptyset$ then $\mathbf{P}(E_0) = \{\emptyset\}$ and so $|\mathbf{P}(E_0)| = 1 = 2^0$. Now assume that $n \in S$. We have to show that $n + 1 \in S$. Suppose then that E_{n+1} is a set with $|E_{n+1}| = n + 1$ and let $x \in E_{n+1}$. If $X = E_{n+1} \setminus \{x\}$ then $|X| = n$ and so $|\mathbf{P}(X)| = 2^n$. But

$$\mathbf{P}(E_{n+1}) = \mathbf{P}(X) \cup \{A \cup \{x\} \; ; \; A \in \mathbf{P}(X)\},$$

and since

$$\mathbf{P}(X) \cap \{A \cup \{x\} \; ; \; A \in \mathbf{P}(X)\} = \emptyset,$$

we can apply 1.13 to obtain

$$\begin{aligned}
|\mathbf{P}(E_{n+1})| &= |\mathbf{P}(X)| + |\{A \cup \{x\} \; ; \; A \in \mathbf{P}(X)| \\
&= 2^n + 2^n \\
&= 2^{n+1}.
\end{aligned}$$

Thus $n + 1 \in S$ and so $S = \mathbb{N}$ as required.

For some applications the following restatement of 4.1 is useful.

4.2 Theorem [Second principle of induction] *Given* $m \in \mathbb{N}$, *let*

$$S_m = \{x \in \mathbb{N} \; ; \; x \geq m\}.$$

If S *is a subset of* S_m *such that*

(1) $m \in S$;

(2) $t \in S$ *for* $m \leq t \leq n$ *implies* $n + 1 \in S$,

then we have $S = S_m$.

Proof This is an immediate consequence of 4.1. $\Diamond$

Example Define a sequence $(a_n)_{n \geq 0}$ by

$$a_0 = 1, a_1 = 2, \text{ and } (\forall n \geq 2) \; a_n = 4a_{n-1} - 4a_{n-2}.$$

We show, using the second principle of induction, that $a_n = 2^n$ for every $n \in \mathbb{N}$. Let

$$S = \{n \in \mathbb{N} \; ; \; a_n = 2^n\}.$$

Note that $0 \in S$ since $a_0 = 1 = 2^0$. Next, $1 \in S$ since $a_1 = 2 = 2^1$. Now suppose that $t \in S$ for $0 \leq t \leq n$ where $n \geq 1$. We show that this implies that $n + 1 \in S$. We have $a_n = 2^n, a_{n-1} = 2^{n-1}$ and so

$$
\begin{aligned}
a_{n+1} &= 4a_n - 4a_{n-1} \\
&= 4 \cdot 2^n - 4 \cdot 2^{n-1} \\
&= 2^{n+2} - 2^{n+1} \\
&= 2^{n+1}(2 - 1) \\
&= 2^{n+1},
\end{aligned}
$$

as required.

Note that we had to check the case $n = 1$ separately in this Example. The most common error in applying induction results from a failure to check cases not covered by the general result. We illustrate this as follows.

Example Define a sequence $(a_n)_{n \geq 0}$ by

$$a_0 = 0, a_1 = \alpha, \text{ and } (\forall n \geq 2) \; a_n = 2a_{n-1} - a_{n-2}.$$

Then we have, for all $n \in \mathbb{N}$, $a_n = n$ if $\alpha = 1$, and $a_n = 0$ if $\alpha = 0$. In fact, suppose first that $\alpha = 1$ and let

$$S = \{n \in \mathbb{N} \; ; \; a_n = n\}.$$

Clearly, $0 \in S$ and $1 \in S$. Consider (2) of 4.2 for $n \geq 1$. If $a_n = n$ and $a_{n-1} = n - 1$ then

$$a_{n+1} = 2a_n - a_{n-1} = 2n - (n - 1) = n + 1.$$

So if $t \in S$ for $0 \leq t \leq n$ then we have $n + 1 \in S$. Thus $S = \mathbb{N}$ and $a_n = n$ for all $n \in \mathbb{N}$. A similar induction shows that if $\alpha = 0$ then $a_n = 0$ for all $n \in \mathbb{N}$.

Definition If $r, n \in \mathbb{N}$ are such that $0 \leq r \leq n$ define the *binomial coefficient* $\binom{n}{r}$ by

$$\binom{n}{r} = \frac{n!}{r!(n - r)!}.$$

4.3 Theorem (1) *For $1 \leq r \leq n$ we have*

$$\binom{n+1}{r} = \binom{n}{r} + \binom{n}{r-1}.$$

(2) *For $0 \leq r \leq n$ we have $\binom{n}{r} \in \mathbb{N}$.*

Proof (1) Using the above definition, we see that

$$\binom{n}{r} + \binom{n}{r-1} = \frac{n!}{r!(n-r)!} + \frac{n!}{(r-1)!(n-r+1)!}$$

$$= \frac{(n-r+1)n! + r \cdot n!}{r!(n-r+1)!}$$

$$= \frac{(n+1)!}{r!(n+1-r)!}$$

$$= \binom{n+1}{r}.$$

(2) First note that if $n = 0$ then there is only one binomial coefficient $\binom{0}{0} = 1 \in \mathbb{N}$. Also, if $n = 1$ then there are two binomial coefficients, namely

$$\binom{1}{0} = 1 \in \mathbb{N} \quad \text{and} \quad \binom{1}{1} = 1 \in \mathbb{N}.$$

Suppose now that $n \geq 1$ and that $\binom{t}{r} \in \mathbb{N}$ for $r \leq t \leq n$. We can complete the proof using 4.2 if we can show that $\binom{n+1}{r} \in \mathbb{N}$ for $0 \leq r \leq n+1$. But

$$\binom{n+1}{0} = 1 \in \mathbb{N} \quad \text{and} \quad \binom{n+1}{n+1} = 1 \in \mathbb{N},$$

while for $1 \leq r \leq n$ we have, by (1),

$$\binom{n+1}{r} = \binom{n}{r} + \binom{n}{r-1}.$$

Then $\binom{n}{r} \in \mathbb{N}$ and $\binom{n}{r-1} \in \mathbb{N}$ give $\binom{n+1}{r} \in \mathbb{N}$ as required. $\diamond$

Example Let E be a set with $|E| = n$. Then E contains $\binom{n}{r}$ subsets A with $|A| = r$. We prove this by induction. The result is clearly true for $n = 0$ and $n = 1$. Suppose that it holds for all sets having no more than n elements where $n \geq 1$. We show that if E is a set with $n + 1$ elements then E contains $\binom{n+1}{r}$ subsets A with $|A| = r$. Observe that the result holds for $r = n + 1$ and $r = 0$, so suppose that $1 \leq r \leq n$. Let $x \in E$ and consider $E \setminus \{x\}$. By hypothesis, $E \setminus \{x\}$ contains $\binom{n}{r}$ subsets A_i with $|A_i| = r$, and $\binom{n}{r-1}$ subsets B_j with $|B_j| = r - 1$. Now the subsets of E containing r elements are precisely the subsets A_i and $B_j \cup \{x\}$, and so E contains

$$\binom{n}{r} + \binom{n}{r-1}$$

subsets with r elements. The result now follows by 4.3(1).

The set $\mathbb{N}$ of natural numbers is a subset of the set $\mathbb{Z}$ of integers. We can extend the natural order on $\mathbb{N}$ to $\mathbb{Z}$. For $a, b \in \mathbb{Z}$ we define

$$a \leq b \iff b - a \in \mathbb{N}.$$

Again, if $a \leq b$ and $a \neq b$ then we write $a < b$.

Definition If $m, n \in \mathbb{Z}$ then we say that m *divides* n (or that m is a *factor* of n) if $n = md$ for some $d \in \mathbb{Z}$.

We shall sometimes use the notation $m|n$ to denote the fact that m divides n.

Example Let the sequence $(a_n)_{n \geq 0}$ be given by

$$a_n = 3^{2n+4} - 2^{2n}.$$

We show by induction that $5 | a_n$ for every $n \in \mathbb{N}$. First note that since $a_0 = 80$ we have $5 | a_0$. Suppose then that $5 | a_n$. We prove as follows that $5 | a_{n+1}$. Observe that

$$
\begin{aligned}
a_{n+1} + a_n &= 3^{2n+6} - 2^{2n+2} + 3^{2n+4} - 2^{2n} \\
&= 3^{2n+4}(3^2 + 1) - 2^{2n}(2^2 + 1) \\
&= 10 \cdot 3^{2n+4} - 5 \cdot 2^{2n},
\end{aligned}
$$

and so $5 | (a_{n+1} + a_n)$. Since $5 | a_n$ by hypothesis, it follows that $5 | a_{n+1}$ as required.

Example The relation R defined on $\mathbb{N}$ by

$$mRn \iff m | n$$

is an order on $\mathbb{N}$. But the corresponding relation S defined on $\mathbb{Z}$ by

$$mSn \iff m | n$$

is not an order on $\mathbb{Z}$ since it fails to be anti-symmetric; we have, for example, $2 | -2$ and $-2 | 2$ with $2 \neq -2$.

Divisibility properties of integers are derived from the following fundamental result.

4.4 Theorem [Euclidean division] *If $m, n \in \mathbb{Z}$ with $m \neq 0$ then there are unique integers q and r such that*

$$n = mq + r \text{ with } 0 \leq r < |m|.$$

Proof Consider first the case where $m > 0$. If

$$S = \{ mx - n ; \ x \in \mathbb{Z} \}$$

then S contains positive integers and so, since $\mathbb{N}$ is well ordered, S contains a least positive integer, say $mt - n$. Writing $t = q + 1$, we have

$$m(q + 1) - n > 0 \text{ and } mq - n \leq 0,$$

and so $0 \leq n - mq < m$. Thus

$$n = mq + (n - mq)$$
$$= mq + r,$$

where $r = n - mq$. But $0 \leq r < m$ and so the result holds in this case.

Now suppose that $m < 0$. Then $-m > 0$ and so, by the above, we have $n = (-m)q_1 + r$ where $0 \leq r < -m$. Writing $q = -q_1$, we thus have that $n = mq + r$ with $0 \leq r < |m|$ as required.

It remains to show that the integers q, r are unique. Suppose then that we have

$$n = mq_1 + r_1 = mq_2 + r_2$$

with $0 \leq r_1 < |m|$ and $0 \leq r_2 < |m|$. If $q_1 \neq q_2$ then clearly we have $|m(q_1 - q_2)| > |m|$. But

$$|m(q_1 - q_2)| = |r_1 - r_2| < |m|$$

and so we have a contradiction. We conclude therefore that $q_1 = q_2$, whence it follows that also $r_1 = r_2$. $\Diamond$

Definition The unique integers q and r appearing in 4.4 are called the *quotient* and *remainder* on dividing n by m.

We shall now consider some consequences of euclidean division.

Definition Let m and n be non-zero integers. A positive integer d is called a *highest common factor* of m and n if

(1) $d|m$ and $d|n$;
(2) if $t|m$ and $t|n$ then $t|d$.

Example A highest common factor of 18 and 24 is 6.

Concerning the existence and uniqueness of highest common factors, we have the following result.

4.5 Theorem *If m and n are non-zero integers then*

(1) *a highest common factor d of m, n exists;*
(2) *d is unique;*

(3) *there exist* $u, v \in \mathbb{Z}$ *such that*

$$d = mu + nv.$$

Proof We show first that if d_1 and d_2 are each highest common factors of m, n then $d_1 = d_2$. In fact, we have $d_1|m$ and $d_1|n$ and so, since d_2 is a highest common factor of m and n, we see that d_1 divides d_2. Similarly, d_2 divides d_1. Since $d_1, d_2 \in \mathbb{N}$, it follows that $d_1 = d_2$.

Note now that the proof of (2) will be complete once (1) and (3) have been established. For this purpose, let

$$S = \{mx + ny \; ; \; x, y \in \mathbb{Z}\}.$$

Clearly, S contains positive integers and so, by the well ordering of $\mathbb{N}$, S contains a least positive integer, d say. Then

$$d = mu + nv$$

for some $u, v \in \mathbb{Z}$. We show that d is a highest common factor of m, n and thereby establish both (1) and (3).

By 4.4 we have $m = qd + r$ where $0 \leq r < d$ and so

$$\begin{aligned}
r &= m - qd \\
&= m - q(mu + nv) \\
&= m(1 - qu) + n(-qv).
\end{aligned}$$

This shows that $r \in S$. Since $r < d$ and d is by definition the least positive integer in S, it follows that we must have $r = 0$. Consequently $m = qd$ and so $d|m$. A similar argument shows that $d|n$. Finally, suppose that $t|m$ and $t|n$. Then since $d = mu + nv$ we have that $t|d$. Thus d is a highest common factor of m and n. $\diamondsuit$

Since highest common factors are unique we can talk of *the* highest common factor of m and n. We shall denote this by

$$\mathrm{hcf}\{m, n\}.$$

Example $\mathrm{hcf}\{18, 30\} = 6$. We have $6 = 2 \cdot 18 - 1 \cdot 30$.

Example Note that the integers u and v of 4.5(3) are not unique. For example,

$$6 = 2 \cdot 18 - 1 \cdot 30 = 7 \cdot 18 - 4 \cdot 30 = -3 \cdot 18 + 2 \cdot 30.$$

Although 4.5 guarantees that the highest common factor of two non-zero integers exists, it does not give a practical method of calculating it. We would, at the same time, like a practical method of finding $u, v \in \mathbb{Z}$ such that

$$\mathrm{hcf}\{m, n\} = mu + nv.$$

The next result not only provides a second proof of the existence of highest common factors but also provides these methods.

4.6 Theorem [Division algorithm] *Let m and n be non-zero integers. Consider repeated applications of euclidean division to obtain*

$$
\begin{aligned}
n &= mq_1 + r_1, & 0 < r_1 < |m| \\
m &= r_1 q_2 + r_2, & 0 < r_2 < r_1 \\
r_1 &= r_2 q_3 + r_3, & 0 < r_3 < r_2 \\
&\;\;\vdots \\
r_i &= r_{i+1} q_{i+2} + r_{i+2}, & 0 < r_{i+2} < r_{i+1} \\
&\;\;\vdots \\
r_{t-2} &= r_{t-1} q_t + r_t, & 0 < r_t < r_{t-1} \\
r_{t-1} &= r_t q_{t+1},
\end{aligned}
$$

noting that the process must reach an equation with zero remainder. Then the last non-zero remainder r_t in this process is the highest common factor of m and n. Moreover, these equations may be used in the reverse order, starting with

$$\mathrm{hcf}\{m, n\} = r_t = r_{t-2} - r_{t-1} q_t,$$

to find integers u, v such that $\mathrm{hcf}\{m, n\} = mu + nv$.

Proof Observe that the remainders r_i in the above process form a decreasing chain

$$|m| > r_1 > r_2 > \cdots > r_i > \cdots$$

of positive integers, so there can be no more than $|m| - 1$ non-zero remainders.

To see that $r_t = \text{hcf}\{m, n\}$, observe that the final equation

$$r_{t-1} = r_t q_{t+1}$$

shows that $r_t | r_{t-1}$. The penultimate equation

$$r_{t-2} = r_{t-1} q_t + r_t$$

now shows that $r_t | r_{t-2}$. Continuing in this way up the above chain of equations, we see that $r_t | m$ and $r_t | n$. Suppose now that k is such that $k | m$ and $k | n$. Then from the first equation

$$n = mq_1 + r_1$$

we see that $k | r_1$. The second equation

$$m = r_1 q_2 + r_2$$

now shows that $k | r_2$. Continuing in this way down the above chain of equations, we see that $k | r_t$. Thus we see that r_t is the highest common factor of m and n.

To find integers u, v such that $r_t = mu + nv$, we again use the equations in reverse order to obtain

$$\begin{aligned}
r_t &= r_{t-2} - r_{t-1} q_t \\
&= r_{t-2} - (r_{t-3} - r_{t-2} q_{t-1}) q_t \\
&= -r_{t-3} q_t + r_{t-2}(1 + q_{t-1} q_t) \\
&\;\;\vdots \\
&= mu + nv. \qquad \diamond
\end{aligned}$$

Example To find the highest common factor d of 26 and 382, and integers u, v such that $d = 26u + 382v$, we compute the chain of equations as in 4.6. We obtain

$$\begin{aligned}
382 &= 26 \cdot 14 + 18 \\
26 &= 18 \cdot 1 + 8 \\
18 &= 8 \cdot 2 + 2 \\
8 &= 2 \cdot 4,
\end{aligned}$$

from which we see, by 4.6, that $\mathrm{hcf}\{26, 382\} = 2$. Also,

$$
\begin{aligned}
2 &= 18 - 8 \cdot 2 \\
&= 18 - (26 - 18 \cdot 1) \cdot 2 \\
&= 18 \cdot 3 - 26 \cdot 2 \\
&= (382 - 26 \cdot 14) \cdot 3 - 26 \cdot 2 \\
&= 382 \cdot 3 - 26 \cdot 44,
\end{aligned}
$$

and so $u = -44$ and $v = 3$ are such that

$$
2 = \mathrm{hcf}\{26, 382\} = 26u + 382v.
$$

Example The number of steps in the division algorithm can be reduced using the following method. Write

$$
n = mq_1 \pm r_1, \qquad 0 < r_1 < |m|,
$$

choosing the positive or negative sign to minimize the remainder r_1. Similarly, at each stage

$$
r_i = r_{i+1}q_{i+2} \pm r_{i+2}, \qquad 0 < r_{i+2} < r_{i+1}
$$

choose the sign to minimize the remainder. We leave to the reader the easy exercise of showing that in so doing the proof of 4.6 remains valid. By way of example, to find again the highest common factor of 26 and 382, we proceed as follows.

$$
\begin{aligned}
382 &= 26 \cdot 15 - 8 \\
26 &= 8 \cdot 3 + 2 \\
8 &= 2 \cdot 4,
\end{aligned}
$$

so $\mathrm{hcf}\{26, 382\} = 2$. Also,

$$
\begin{aligned}
2 &= 26 - 8 \cdot 3 \\
&= 26 - (26 \cdot 15 - 382) \cdot 3 \\
&= 26 \cdot (-44) + 382 \cdot 3.
\end{aligned}
$$

We now consider the notion that is dual to that of a highest common factor.

Definition Let m and n be non-zero integers. A positive integer ℓ is said to be a *least common multiple* of m and n if

(1) $m|\ell$ and $n|\ell$;

(2) if $m|t$ and $n|t$ then $\ell|t$.

Example A least common multiple of 18 and 24 is 72.

Example Let k be the least positive integer such that $m|k$ and $n|k$. Then k is a least common multiple of m and n. To see this, syppose that $m|t$ and $n|t$. We show as follows that $k|t$. By euclidean division, we have

$$t = kq + r, \qquad 0 \leq r < k.$$

But since $m|k$ and $m|t$ we have $m|r$; and similarly, $n|r$. If now $r > 0$ then this contradicts the minimality of k. Thus $r = 0$ and $k|t$ as required.

A proof similar to that of the uniqueness of highest common factors shows that least common multiples are also unique. We shall denote the least common multiple of m and n by $\mathrm{lcm}\{m, n\}$.

Definition An integer p is said to be *prime* if

(1) $p > 1$,

(2) the only positive divisors of p are 1 and p.

Example Every positive integer greater than 1 has a prime factor. To see this,we use the second principle of induction. Clearly, 2 has the prime factor 2. Suppose now that n is such that every integer t with $2 \leq t \leq n$ has a prime factor. We show that this implies that $n + 1$ has a prime factor. Now if $n + 1$ has has no factor s with $2 \leq s < n + 1$ then $n + 1$ is a prime and so has itself as a prime factor. If however $n + 1$ does have a factor s with $2 \leq s < n + 1$ then $2 \leq s \leq n$ and so, by the induction hypothesis, s has a prime factor p. Since p divides s and s divides $n + 1$ it follows that p divides $n + 1$ as required.

Example Every positive integer greater than 1 can be written as a product of prime factors. To show this, we again use the second principle of induction. It is clear that 2, being itself a prime, can be written trivially as a product of primes. Suppose now that n is such that every integer t with $2 \leq t \leq n$ can be written as a product of primes. By the previous Example, $n + 1$ has a prime factor, p say. If $p = n + 1$ then $n + 1$ is a product of primes and we are done. Otherwise, $p < n + 1$ and so

$$2 \le \frac{n + 1}{p} \le n.$$

Then, by the induction hypothesis, we have

$$\frac{n + 1}{p} = p_1 p_2 \cdots p_k$$

for primes $p_1, \ldots, p_k$. It follows that $n + 1$ is a product of primes as required.

4.7 Theorem [Euclid] $P = \{p \in \mathbb{N} \; ; \; p \text{ is prime}\}$ *is infinite.*

Proof Suppose that P were finite, say $P = \{p_1, \ldots, p_n\}$, and consider the integer

$$k = 1 + p_1 p_2 \cdots p_n.$$

Observe that none of the p_i divide k. But we have proved above that every integer greater than 1 has a prime factor. This contradiction shows that P must be infinite. $\diamond$

Our objective now is to prove that the 'prime decomposition' given in the preceding Example is essentially unique. By way of preparation for this, we note the following results.

4.8 Theorem *Suppose that $a, b \in \mathbb{Z}$ and let p be a prime such that $p|ab$. Then either $p|a$ or $p|b$.*

Proof Suppose that p does not divide a. We show as follows that p must divide b. Since p does not divide a we have $\mathrm{hcf}\{p, a\} = 1$ and so, by 4.5, there are integers u, v such that

$$1 = pu + av.$$

It follows that $b = pub + abv$ and hence $p|b$ since, by hypothesis, $p|ab$. $\diamond$

4.9 Corollary *Let $a_1, \ldots, a_n$ be integers and let p be a prime such that $p|a_1 a_2 \cdots a_n$. Then $p|a_i$ for some i.*

Proof We use induction on n. The result clearly holds if $n = 1$, and the case when $n = 2$ is established by 4.8. Suppose now that the result holds for all products of n integers with $n \geq 2$, and suppose that

$$p | a_1 a_2 \cdots a_{n+1}.$$

Let $b = a_2 a_3 \cdots a_{n+1}$. Then we have $p | a_1 b$ and so, by 4.8, either $p | a_1$ or $p | b$. If $p | a_1$ the proof is complete; and if $p | b$ then, by the induction hypothesis, $p | a_i$ for some i with $2 \leq i \leq n + 1$. $\diamond$

4.10 Theorem *Let n be an integer greater than 1. Then n has an expression as a product of primes which is unique up to the order in which the prime factors occur.*

Proof We have already seen that n can be expressed as a product

$$n = p_1 p_2 \cdots p_r$$

where $p_1, \ldots, p_r$ are primes with $p_1 \leq p_2 \leq \cdots \leq p_r$. Suppose that we also have

$$n = q_1 q_2 \cdots q_s$$

where $q_1, \ldots, q_s$ are primes with $q_1 \leq q_2 \leq \cdots \leq q_s$. We prove as follows that $r = s$ and $p_i = q_i$ for $i = 1, \ldots, r$.

The result is certainly true for $n = 2$. Assume by way of induction that it holds for every integer t with $2 \leq t \leq n - 1$ and consider the decompositions

$$n = p_1 p_2 \cdots p_r = q_1 q_2 \cdots q_s.$$

Observe that $p_1 | q_1 q_2 \cdots q_s$ and so, by 4.9, $p_1 | q_i$ for some i with $1 \leq i \leq s$; and since q_i is also a prime we must then have $p_1 = q_i$. Likewise, $q_1 | p_1 p_2 \cdots p_r$ and so $q_1 | p_j$ for some j with $1 \leq j \leq r$; and since p_j is also prime, $q_1 = p_j$. We now have

$$p_1 = q_i \geq q_1 = p_j \geq p_1,$$

from which we deduce that $p_1 = q_1$. Writing $n = p_1 m$ we then have

$$m = p_2 \cdots p_r = q_2 \cdots q_s$$

and so, by the induction hypothesis, $r = s$ and $p_i = q_i$ for each i. The result now follows. $\diamond$

4.11 Corollary *Let n be an integer greater than 1. Then n has a unique expression in the form*

$$n = p_1^{\alpha_1} p_2^{\alpha_2} \cdots p_r^{\alpha_r}$$

where $p_1, \ldots, p_r$ are primes with $p_1 < p_2 < \cdots < p_r$ and each α_i is a positive integer.

Proof This is an immediate consequence of 4.10. $\diamond$

It is sometimes useful to consider prime decompositions as described in 4.11 in which some of the powers α_i are allowed to be zero. The following illustrates this point.

Example Let m and n be non-zero integers and let $p_1, \ldots, p_r$ be the distinct prime factors of m or n. Then we can write

$$m = p_1^{\alpha_1} p_2^{\alpha_2} \cdots p_r^{\alpha_r}, \quad n = p_1^{\beta_1} p_2^{\beta_2} \cdots p_r^{\beta_r}$$

where $\alpha_i \geq 0$ and $\beta_i \geq 0$ for each i. For example,

$$45 = 2^0 \cdot 3^2 \cdot 5, \qquad 50 = 2^1 \cdot 3^0 \cdot 5^2.$$

If we define

$$\gamma_i = \min\{\alpha_i, \beta_i\}, \quad \delta_i = \max\{\alpha_i, \beta_i\}$$

then it is easy to see that

$$\mathrm{hcf}\{m, n\} = p_1^{\gamma_1} p_2^{\gamma_2} \cdots p_r^{\gamma_r}$$
$$\mathrm{lcm}\{m, n\} = p_1^{\delta_1} p_2^{\delta_2} \cdots p_r^{\delta_r}.$$

Since $\alpha_i + \beta_i = \gamma_i + \delta_i$ for each i, it follows that

$$\mathrm{hcf}\{m, n\} \, \mathrm{lcm}\{m, n\} = mn.$$

Definition If m and n are non-zero integers then we say that m and n are *coprime* if $\mathrm{hcf}\{m, n\} = 1$.

Note that m and n are coprime if and only if no prime p occurs in the prime decompositions of both m and n. Also, it follows from the identity established at the end of the preceding Example that m and n are coprime if and only if $\mathrm{lcm}\{m, n\} = mn$.

Definition Let $\mathbb{Z}^+ = \{n \in \mathbb{Z} \; ; \; n > 0\}$. Then the *Euler φ-function* is the mapping $\varphi : \mathbb{Z}^+ \to \mathbb{N}$ given by

$$\varphi(n) = |\{t \in \mathbb{N} \; ; \; 1 \le t \le n, \; \mathrm{hcf}\{t, n\} = 1\}|.$$

The Euler φ-function counts the number of coprime positive integers that are less than or equal to a given positive integer n. Thus, for example,

$$\varphi(1) = 1, \; \varphi(2) = 1, \; \varphi(3) = 2, \; \varphi(4) = 2, \; \varphi(5) = 4, \; \cdots.$$

4.12 Theorem *Let p be a prime and let k be a positive integer. Then*

$$\varphi(p^k) = p^k - p^{k-1}.$$

Proof Let $E = \{1, 2, \ldots, p^k\}$. Then

$$\varphi(p^k) = |\{t \in E \; ; \; t \text{ is coprime to } p\}|.$$

Observe that $|E| = p^k$ and so we have

$$\varphi(p^k) = p^k - |\{t \in E \; ; \; p|t\}|.$$

But if $p|t$ then $t = pn$ where necessarily

$$n \in \{1, 2, \ldots, p^{k-1}\}.$$

It follows that $|\{t \in E \; ; \; p|t\}| = p^{k-1}$ and hence

$$\varphi(p^k) = p^k - p^{k-1}. \quad \Diamond$$

Our objective now is to determine an explicit formula for $\varphi(n)$. In order to do so, we shall consider linear congruences.

It is easy to solve a linear equation $mx + n = 0$ over $\mathbb{Z}$. If we are given $a, b \in \mathbb{Z}$ then the equation $ax = b$ has a solution in $\mathbb{Z}$ if and only if $a|b$. The corresponding problem where equality is replaced by the equivalence relation $\equiv_m$ is more interesting and gives rise to the following notion.

Definition By a *linear congruence* we shall mean an expression of the form

$$ax \equiv b(\mathrm{mod} \; m)$$

in which $a, b, m \in \mathbb{Z}$ with m positive.

4.13 Theorem *Suppose that $a, b, m \in \mathbb{Z}$ with m positive. Then if the linear congruence*

$$ax \equiv b(\text{mod } m)$$

has a solution α then every $\beta \in [\alpha]_m$ is also a solution.

Proof Since α is a solution of the congruence, there exists $\lambda \in \mathbb{Z}$ such that

$$a\alpha - b = \lambda m.$$

But if $\beta \in [\alpha]_m$ then $\beta = \alpha + \mu m$ for some $\mu \in \mathbb{Z}$ and so

$$\lambda m = a\alpha - b = a(\beta - \mu m) - b = a\beta - a\mu m - b.$$

Then $a\beta - b = (\lambda - a\mu)m$ and so $a\beta \equiv b(\text{mod } m)$. $\diamond$

Recall that the equivalence classes $[n]_m$ partition $\mathbb{Z}$ and that there are precisely m distinct classes. If we let

$$\mathbb{Z}_m = \{0, 1, \ldots, m - 1\},$$

then each equivalence class contains an element of $\mathbb{Z}_m$, and no two elements of $\mathbb{Z}_m$ belong to the same class. Therefore, using 4.13, we see that if the linear congruence

$$ax \equiv b(\text{mod } m)$$

has a solution α then it has a solution $\beta \in \mathbb{Z}_m$ where $\beta \equiv \alpha(\text{mod } m)$. In other words, the linear congruence is completely solved once we have determined the solutions in $\mathbb{Z}_m$.

Example The linear congruence

$$2x \equiv 1(\text{mod } 4)$$

has no solution. For, considering each of the elements 0,1,2,3 of $\mathbb{Z}_4$ in turn, we have

$$2 \cdot 0 \equiv 0(\text{mod } 4), \quad 2 \cdot 1 \equiv 2(\text{mod } 4),$$
$$2 \cdot 2 \equiv 0(\text{mod } 4), \quad 2 \cdot 3 \equiv 2(\text{mod } 4).$$

However, the linear congruence

$$2x \equiv 2(\text{mod } 8)$$

has solutions $1, 5 \in \mathbb{Z}_8$; and the linear congruence

$$4x \equiv 4(\text{mod } 8)$$

has solutions $1, 3, 5, 7 \in \mathbb{Z}_8$.

We shall now determine precisely when a linear congruence has a solution, and the number of solutions that exist.

4.14 Theorem *Let* $a, b, m \in \mathbb{Z}$ *with* m *positive and let* $d = \mathrm{hcf}\{a, m\}$. *Then the linear congruence*

$$(1) \qquad\qquad ax \equiv b(\mathrm{mod}\ m)$$

has a solution if and only if $d|b$, *in which case the number of solutions in* $\mathbb{Z}_m$ *is* d.

Proof Suppose that $d|b$, say $b = dk$ for some $k \in \mathbb{Z}$. By 4.5 there exist $u, v \in \mathbb{Z}$ such that

$$d = au + mv,$$

so $b = dk = auk + mvk$. It then follows that

$$b \equiv auk(\mathrm{mod}\ m),$$

so that (1) has the solution uk.

Conversely, suppose that $\alpha \in \mathbb{Z}$ is a solution of (1). Then for some $\lambda \in \mathbb{Z}$ we have

$$a\alpha - b = \lambda m.$$

Then $b = a\alpha - \lambda m$ and so, since $d|a$ and $d|m$, we have $d|b$ as required.

In order to prove that (1) has d solutions in $\mathbb{Z}_m$, we first prove that if $d = 1$ then (1) has a unique solution in $\mathbb{Z}_m$. In fact, if in this case $\alpha \in \mathbb{Z}_m$ and $\beta \in \mathbb{Z}_m$ are solutions then from

$$a\alpha \equiv b(\mathrm{mod}\ m), \quad a\beta \equiv b(\mathrm{mod}\ m)$$

we have $a(\alpha - \beta) \equiv 0(\mathrm{mod}\ m)$. Since by hypothesis $\mathrm{hcf}\{a, m\} = 1$, it follows that $m|(\alpha - \beta)$. But no two elements of $\mathbb{Z}_m$ belong to the same class modulo m and so we deduce that $\alpha = \beta$.

Now suppose that $d > 1$ and that $d|b$. Write

$$a = a_1 d, \quad b = b_1 d, \quad m = m_1 d.$$

Observe that

$$ax \equiv b(\mathrm{mod}\ m) \iff a_1 x \equiv b_1(\mathrm{mod}\ m_1).$$

Now $\mathrm{hcf}\{a_1, m_1\} = 1$ and so, by the previous observation, the congruence $a_1 x \equiv b_1 (\bmod\ m_1)$ has a unique solution $\alpha \in \mathbb{Z}_{m_1}$. The result now follows from the fact that the solutions of (1) in $\mathbb{Z}_m$ are precisely

$$\alpha, \quad \alpha + m_1, \quad \alpha + 2m_1, \quad \ldots, \alpha + (d-1)m_1. \quad \diamondsuit$$

4.15 Corollary *If $a, b, m \in \mathbb{Z}$ with m positive then the linear congruence*

$$ax \equiv b(\bmod\ m)$$

has a unique solution in $\mathbb{Z}_m$ if and only if $\mathrm{hcf}\{a, m\} = 1$. $\diamondsuit$

Example We use the method of 4.14 to solve the linear congruence

$$8x \equiv 4(\bmod\ 12).$$

Since $\mathrm{hcf}\{8, 12\} = 4$, we see that the congruence has exactly 4 solutions in $\mathbb{Z}_{12}$. In this case it would be a simple matter to check each of the twelve elements in turn. However, we follow the method of 4.14. First we find a solution of

$$2x \equiv 1(\bmod\ 3)$$

in $\mathbb{Z}_3$. Clearly, 2 is such a solution. The given congruence

$$8x \equiv 4(\bmod\ 12)$$

then has the four solutions

$$2, \quad 2+3, \quad 2+6, \quad 2+9,$$

i.e. $2, 5, 8, 11 \in \mathbb{Z}_{12}$.

For larger integers we have to do slightly more work, as the next Example shows.

Example To find all the solutions in $\mathbb{Z}_{385}$ of the linear congruence

$$224x \equiv 154(\bmod\ 385).$$

First we determine $\mathrm{hcf}\{224, 385\}$ using the division algorithm. We have

$$385 = 224 \cdot 2 - 63$$
$$224 = 63 \cdot 4 - 28$$
$$63 = 28 \cdot 2 + 7$$
$$28 = 4 \cdot 7,$$

and so $\mathrm{hcf}\{224, 385\} = 7$. Since $7|154$, the given congruence has 7 solutions in $\mathbb{Z}_{385}$. The above equations give

$$7 = 63 - 28 \cdot 2$$
$$= 224 \cdot 2 - 63 \cdot 7$$
$$= 385 \cdot 7 - 224 \cdot 12,$$

from which we obtain $1 = 55 \cdot 7 - 32 \cdot 12$. To solve

$$32x \equiv 22(\mathrm{mod}\ 55)$$

we need only multiply this equation by 22 to obtain

$$-32 \cdot 12 \cdot 22 \equiv 22(\mathrm{mod}\ 55).$$

But $-22 \cdot 12 = -264 \equiv 11(\mathrm{mod}\ 55)$, and so the solutions of the linear congruence

$$224x \equiv 154(\mathrm{mod}\ 385)$$

in $\mathbb{Z}_{385}$ are

$$11, \quad 66, \quad 121, \quad 176, \quad 231, \quad 286, \quad 341.$$

Example Suppose that p is a prime and that $a \in \mathbb{Z}_p \setminus \{0\}$. Then we have $\mathrm{hcf}\{a, p\} = 1$ and so, by 4.15, the linear congruence

$$ax \equiv 1(\mathrm{mod}\ p)$$

has a unique solution in $\mathbb{Z}_p$.

We now observe that if $a \equiv a'(\mathrm{mod}\ m)$ and $b \equiv b'(\mathrm{mod}\ m)$ then the linear congruence

$$ax \equiv b(\mathrm{mod}\ m)$$

has the same solutions as the linear congruence

$$a'x \equiv b'(\mathrm{mod}\ m).$$

For, suppose that $a = a' + \lambda m$ and $b = b' + \mu m$ for $\lambda, \mu \in \mathbb{Z}$. Then if α is a solution of $ax \equiv b(\text{mod } m)$ we have $a\alpha = b + \nu m$ for some $\nu \in \mathbb{Z}$. Then

$$a'\alpha = (a - \lambda m)\alpha$$
$$= a\alpha - \lambda m\alpha$$
$$= b + \nu m - \lambda m\alpha$$
$$= b' + m(\mu + \nu - \lambda\alpha),$$

and so $a'\alpha \equiv b'(\text{mod } m)$.

It follows that we need study only linear congruences

$$ax \equiv b(\text{mod } m)$$

where $a, b \in \mathbb{Z}_m$.

Observe that if $x \equiv y(\text{mod } m)$ then for every $a \in \mathbb{Z}$ we have $xa \equiv ya(\text{mod } m)$; for $x-y = km$ gives $xa-ya = (x-y)a = kam$. Thus, solving a linear congruence modulo m is equivalent to solving a linear equation in $\mathbb{Z}_m$.

We shall now examine simultaneous linear congruences associated with coprime moduli. We note first that if $[x]_n$ denotes the 'mod n'-class of x then whenever $\text{hcf}\{m, n\} = 1$ we have

$$[x]_m \cap [x]_n = [x]_{mn}.$$

To see this, observe that if $y \in [x]_{mn}$ then, for some $\lambda \in \mathbb{Z}$,

$$y = x + \lambda mn$$

and so $y \equiv x(\text{mod } m)$ and $y \equiv x(\text{mod } n)$, whence

$$[x]_{mn} \subseteq [x]_m \cap [x]_n.$$

As for the reverse inclusion, if $y \in [x]_m \cap [x]_n$ then, for some $\lambda, \mu \in \mathbb{Z}$,

$$y = x + \lambda m = x + \mu n$$

and so $\lambda m = \mu n$. Now since $\text{hcf}\{m, n\} = 1$ we have $n|\lambda$. Then $\lambda = n\lambda_1$ gives $y = x + \lambda_1 nm$ and so $y \in [x]_{mn}$.

It now follows by 4.13 that if α is a solution to each of the linear congruences

$$a_1 x \equiv b_1(\text{mod } m),$$
$$a_2 x \equiv b_2(\text{mod } n),$$

then every $\beta \in [\alpha]_{mn}$ is also a solution to each of these congruences.

4.16 Theorem *Let m and n be coprime positive integers. If each of the linear congruences*

$$a_1 x \equiv b_1 (\text{mod } m)$$
$$a_2 x \equiv b_2 (\text{mod } n)$$

has a solution then there is in $\mathbb{Z}_{mn}$ a common solution to the pair of congruences.

Proof First we show that the simultaneous linear congruences

$$x \equiv \alpha_1 (\text{mod } m), \quad x \equiv \alpha_2 (\text{mod } n)$$

have a solution in $\mathbb{Z}_{mn}$. Since $\text{hcf}\{m, n\} = 1$ there exist integers u, v such that $mu + nv = 1$. Then $nv \equiv 1 (\text{mod } m)$ and $mu \equiv 1 (\text{mod } n)$. Now observe that $\alpha_1 nv + \alpha_2 mu$ is a solution to each of the congruences; for

$$\alpha_1 nv + \alpha_2 mu \equiv \alpha_1 nv (\text{mod } m)$$
$$\equiv \alpha_1 (\text{mod } m),$$
$$\alpha_1 nv + \alpha_2 mu \equiv \alpha_2 mu (\text{mod } n)$$
$$\equiv \alpha_2 (\text{mod } n).$$

Suppose now that α_1 is a solution to the linear congruence $a_1 x \equiv b_1 (\text{mod } m)$, and that α_2 is a solution to the linear congruence $a_2 x \equiv b_2 (\text{mod } n)$. By the above observation, there is a solution to the simultaneous congruences

$$x \equiv \alpha_1 (\text{mod } m), \quad x \equiv \alpha_2 (\text{mod } n)$$

which is therefore a solution α to the original system. Since

$$[\alpha]_m \cap [\alpha]_n = [\alpha]_{mn},$$

there is a solution $\beta \equiv \alpha (\text{mod } mn)$ with $\beta \in \mathbb{Z}_{mn}$. $\diamond$

4.17 Corollary *Let m, n be coprime positive integers. Then the simultaneous congruences*

$$x \equiv b_1 (\text{mod } m)$$
$$x \equiv b_2 (\text{mod } n)$$

have a unique common solution in $\mathbb{Z}_{mn}$.

Proof That the congruences have a common solution $\alpha \in \mathbb{Z}_{mn}$ follows from 4.16. Suppose that $\beta \in \mathbb{Z}_{mn}$ is also a solution. Then we have $\alpha \equiv \beta \pmod{m}$ and $\alpha \equiv \beta \pmod{n}$ so $m|(\alpha - \beta)$ and $n|(\alpha - \beta)$. Since $\mathrm{hcf}\{m, n\} = 1$, we must have $mn|(\alpha - \beta)$ and hence $\alpha \equiv \beta \pmod{mn}$ whence $\alpha = \beta$. $\diamondsuit$

4.18 Theorem [Chinese remainder] *Let $n_1, \ldots, n_k$ be positive integers that are pairwise coprime. If each of the linear congruences*

$$(i = 1, \ldots, k) \qquad a_i x \equiv b_i \pmod{n_i}$$

has a solution then there is a solution in $\mathbb{Z}_{n_1 n_2 \cdots n_k}$ to the system of k simultaneous linear congruences.

Proof We use induction on k, the number of congruences in the system. If $k = 1$ there is nothing to prove. If $k = 2$ the result follows from 4.16. Suppose now that the result holds for $k - 1$ simultaneous linear congruences, and consider a system of k linear congruences. By the induction hypothesis, the first $k - 1$ congruences have a common solution α, and any solution of

$$x \equiv \alpha \pmod{n_1 n_2 \cdots n_{k-1}}$$

is a solution of the first $k - 1$ congruences. Observe that

$$\mathrm{hcf}\{n_1 n_2 \cdots n_{k-1}, n_k\} = 1$$

and so we can apply 4.16 to the system

$$x \equiv \alpha \pmod{n_1 n_2 \cdots n_{k-1}}$$
$$a_k x \equiv b_k \pmod{n_k}$$

to obtain the result. $\diamondsuit$

Example Consider the simultaneous linear congrueces

$$8x \equiv 4 \pmod{12},$$
$$5x \equiv 10 \pmod{25}.$$

The first congruence has solution set $\{2, 5, 8, 11\}$ in $\mathbb{Z}_{12}$, and the second has solution set $\{2, 7, 12, 17, 22\}$ in $\mathbb{Z}_{25}$. There are therefore $4 \times 5 = 20$ solutions in $\mathbb{Z}_{300}$ that are common to the two congruences. Now 2 belongs to each solution set and so the solution set of the system can be written as

$$\{2 + 15\lambda \in \mathbb{Z}_{300} \; ; \; 0 \leq \lambda \leq 19\}.$$

We end the present Chapter by applying the Chinese Remainder Theorem to obtain further information concerning the Euler φ-function. Recall that in 4.12 we proved that, for every prime p,

$$\varphi(p^k) = p^k - p^{k-1}.$$

Since, by 4.11, every integer greater than 1 can be written as a product of powers of primes, the value of $\varphi(n)$ will be determined once we have been able to compute $\varphi(ab)$ for coprime integers a, b.

4.19 Theorem *If a, b are coprime positive integers then*

$$\varphi(ab) = \varphi(a)\varphi(b).$$

Proof Let $A = \{\alpha_1, \ldots, \alpha_{\varphi(a)}\} \subseteq \mathbb{Z}_a$ be the set of positive integers coprime to a, let $B = \{\beta_1, \ldots, \beta_{\varphi(b)}\} \subseteq \mathbb{Z}_b$ be the set of positive integers coprime to b, and let $C = \{\gamma_1, \ldots, \gamma_{\varphi(ab)}\} \subseteq \mathbb{Z}_{ab}$ be the set of positive integers coprime to ab. To prove that $\varphi(ab) = \varphi(a)\varphi(b)$, it suffices to show that

$$|A \times B| = |C|.$$

This we do by establishing a bijection

$$f : A \times B \to C.$$

We define f as follows : given $(\alpha_i, \beta_j) \in A \times B$, the linear congruences

$$x \equiv \alpha_i (\mathrm{mod} \; a)$$
$$x \equiv \beta_j (\mathrm{mod} \; b)$$

have, by 4.18, a unique common solution $\gamma_k \in \mathbb{Z}_{ab}$. Let

$$d = \mathrm{hcf}\{\gamma_k, ab\}.$$

Since a and b are coprime and $d|ab$, we have that either $d|a$ or $d|b$. Also, $d|\gamma_k$ so either $d|\alpha_i, d|a$ or $d|\beta_j, d|b$. It follows from the fact that α_i and a are coprime, as are β_j and b, that $d = 1$. This then shows that $\gamma_k \in C$ and we can define

$$f(\alpha_i, \beta_j) = \gamma_k.$$

To show that f is injective, suppose that

$$\gamma_k = f(\alpha_i, \beta_j) = f(\alpha_p, \beta_q).$$

Since $\alpha_i \equiv \gamma_k \pmod{a}$ and $\alpha_p \equiv \gamma_k \pmod{a}$ we have $\alpha_i \equiv \alpha_p \pmod{a}$. Since $\alpha_i, \alpha_p \in \mathbb{Z}_a$ it follows that $\alpha_i = \alpha_p$. Similarly, $\beta_j = \beta_q$ and hence f is injective.

To show that f is surjective, let $\gamma \in C$. Then since γ is coprime to ab it is clear that γ is coprime to a and to b. By 4.15, the linear congruence

$$x \equiv \gamma \pmod{a}$$

has a solution $\alpha \in \mathbb{Z}_a$, and since $\mathrm{hcf}\{\gamma, a\} = 1$ we must have $\alpha \in A$. Similarly, there is a $\beta \in B$ with $\beta \equiv \gamma \pmod{b}$. But by the definition of f we have

$$f(\alpha, \beta) = \gamma$$

and so f is surjective. Thus f is a bijection as required. $\Diamond$

4.20 Corollary *If $n_1, \ldots, n_k$ are positive integers that are pairwise coprime then*

$$\varphi(n_1 \cdots n_k) = \varphi(n_1) \cdots \varphi(n_k).$$

Proof Use induction on k. The inductive step is provided for by 4.19. $\Diamond$

4.22 Corollary *Let n be an integer greater than 2 and let*

$$n = p_1^{\alpha_1} p_2^{\alpha_2} \cdots p_k^{\alpha_k}$$

be the prime factorisation of n. Then

$$\varphi(n) = n\left(1 - \frac{1}{p_1}\right)\left(1 - \frac{1}{p_2}\right) \cdots \left(1 - \frac{1}{p_k}\right).$$

Proof Observe that for $i = 1, \ldots, k$ we have

$$\varphi(p_i^{\alpha_i}) = p_i^{\alpha_i} - p_i^{\alpha_i - 1}$$
$$= p_i^{\alpha_i}\left(1 - \frac{1}{p_i}\right).$$

The result now follows by 4.20. $\Diamond$

Example To find $\varphi(151200)$, first write

$$151200 = 2^5 \cdot 3^3 \cdot 5^2 \cdot 7.$$

Then we have

$$
\begin{aligned}
\varphi(151200) &= \varphi(2^5 \cdot 3^3 \cdot 5^2 \cdot 7) \\
&= \varphi(2^5)\varphi(3^3)\varphi(5^2)\varphi(7) \\
&= (32 - 16)(27 - 9)(25 - 5)(7 - 1) \\
&= 16 \cdot 18 \cdot 20 \cdot 6 \\
&= 34560.
\end{aligned}
$$

CHAPTER FIVE

Permutations

The most important type of mapping is certainly a bijection. In this Chapter we shall be particularly interested in the set of bijections on a given finite set.

Definition A bijection $\sigma : X \to X$ is called a *permutation* on X.

We shall denote the set of all permutations on the set X by P_X. The following properties of permutations are immediate consequences of the results in Chapter Two.

5.1 Theorem *For every set X,*

(1) $\mathrm{id}_X \in P_X$;
(2) *if $\sigma, \tau \in P_X$ then $\sigma \circ \tau \in P_X$;*
(3) *if $\sigma \in P_X$ then $\sigma^{-1} \in P_X$.* $\diamond$

In the case where X is finite, it turns out that, in order to check that a mapping $\sigma : X \to X$ belongs to P_X, it suffices to check that σ is either an injection or a surjection, as the following result shows.

5.2 Theorem *Let X be a finite set and $\sigma : X \to X$ any mapping. Then*

(1) $\sigma \in P_X$ *if and only if σ is injective;*
(2) $\sigma \in P_X$ *if and only if σ is surjective.*

Proof It suffices to establish sufficiency in each case.

(1) Suppose that σ is an injection. Then $\sigma^+ : X \to \mathrm{Im}\,\sigma$ is a bijection and so $|X| = |\mathrm{Im}\,\sigma|$. Since $\mathrm{Im}\,\sigma \subseteq X$, it follows that

$$|X| = |\mathrm{Im}\,\sigma| \leq |X|$$

and hence that $\operatorname{Im}\sigma = X$, so σ is also a surjection.

(2) Suppose that σ is a surjection. If X is a singleton then clearly $\sigma = \operatorname{id}_X$ and so is also injective. If $|X| \geq 2$ then for $x_1, x_2 \in X$ with $x_1 \neq x_2$ we cannot have $\sigma(x_1) = \sigma(x_2)$ since this would imply that

$$|\operatorname{Im}\sigma| < |X|$$

and therefore contradict the hypothesis that σ is surjective. It follows that σ is also an injection. $\Diamond$

In what follows we shall assume that $X = \{1, 2, \ldots, n\}$ and shall write P_X as P_n. Our next task is to calculate $|P_n|$.

5.3 Theorem $|P_n| = n!$.

Proof In view of 5.2(1), we need only count the number of injections

$$\sigma : X \to X$$

where $X = \{1, 2, \ldots, n\}$. Now for each such mapping we have $\sigma(1) \in X$ and so there are n possible choices for $\sigma(1)$. Next, since $\sigma(2) \neq \sigma(1)$, we have

$$\sigma(2) \in X \setminus \{\sigma(1)\}.$$

But $|X \setminus \{\sigma(1)\}| = n - 1$ and so there are $n - 1$ possible choices for $\sigma(2)$. Continuing in this way, we see that in general

$$\sigma(i) \in X \setminus \{\sigma(1), \ldots, \sigma(i-1)\},$$

and so there are $n - i + 1$ possible choices of $\sigma(i)$. The total number of injections $\sigma : X \to X$ is therefore

$$n \cdot (n - 1) \cdots 2 \cdot 1 = n!$$

as claimed. $\Diamond$

We shall adopt more than one notation for permutations. The first that we shall use is the following. If $\sigma : X \to X$ is a permutation we shall display the effect of σ on the elements $1, \ldots, n$ of X by writing

$$\sigma = \begin{pmatrix} 1 & 2 & 3 & \ldots & n \\ \sigma(1) & \sigma(2) & \sigma(3) & \ldots & \sigma(n) \end{pmatrix}.$$

Example Let p be a prime and let $\mathbb{Z}_p^*$ denote the set of non-zero elements of $\mathbb{Z}_p$. Given $a \in \mathbb{Z}_p^*$, consider the mapping

$$\lambda_a : \mathbb{Z}_p^* \to \mathbb{Z}_p^*$$

given by $\lambda_a(t) = (at)_p$, where $(at)_p$ is the element of $\mathbb{Z}_p^*$ guaranteed by 4.14 which is a solution of the linear congruence

$$x \equiv at(\text{mod } p).$$

Note that $(at)_p \neq 0$ since hcf$\{at, p\} = 1$.

Observe that if

$$at_1 \equiv at_2(\text{mod } p)$$

then $a(t_1 - t_2)$ is divisible by p. Since a is coprime to p, we have $t_1 \equiv t_2(\text{mod } p)$. Therefore

$$\lambda_a(t_1) = \lambda_a(t_2) \Longrightarrow t_1 = t_2$$

and so λ_a is injective. It follows by 5.2 that λ_a is a permutation.

Consider in particular $\lambda_5 : \mathbb{Z}_{11}^* \to \mathbb{Z}_{11}^*$. This is the permutation

$$\lambda_5 = \begin{pmatrix} 1 & 2 & 3 & 4 & 5 & 6 & 7 & 8 & 9 & 10 \\ 5 & 10 & 4 & 9 & 3 & 8 & 2 & 7 & 1 & 6 \end{pmatrix}.$$

As another example, $\lambda_7 : \mathbb{Z}_{11}^* \to \mathbb{Z}_{11}^*$ is

$$\lambda_7 = \begin{pmatrix} 1 & 2 & 3 & 4 & 5 & 6 & 7 & 8 & 9 & 10 \\ 7 & 3 & 10 & 6 & 2 & 9 & 5 & 1 & 8 & 4 \end{pmatrix}.$$

Example Consider the symmetries of an equilateral triangle.

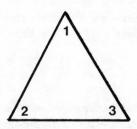

Clockwise rotations through 60° and 120° correspond to the permutations of the vertices described by

$$\begin{pmatrix} 1 & 2 & 3 \\ 3 & 1 & 2 \end{pmatrix}, \quad \begin{pmatrix} 1 & 2 & 3 \\ 2 & 3 & 1 \end{pmatrix}$$

respectively. Reflection in the dotted line corresponds to the permutation of the vertices described by

$$\begin{pmatrix} 1 & 2 & 3 \\ 1 & 3 & 2 \end{pmatrix}.$$

In fact, each of the $3! = 6$ permutations in P_3 corresponds to a symmetry of the equilateral triangle.

Example Consider the symmetries of the square.

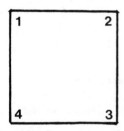

There are $4! = 24$ permutations in P_4 but only eight of these are symmetries of the square, namely

$$\begin{pmatrix} 1 & 2 & 3 & 4 \\ 1 & 2 & 3 & 4 \end{pmatrix}, \quad \begin{pmatrix} 1 & 2 & 3 & 4 \\ 2 & 3 & 4 & 1 \end{pmatrix},$$

$$\begin{pmatrix} 1 & 2 & 3 & 4 \\ 3 & 4 & 1 & 2 \end{pmatrix}, \quad \begin{pmatrix} 1 & 2 & 3 & 4 \\ 4 & 1 & 2 & 3 \end{pmatrix},$$

$$\begin{pmatrix} 1 & 2 & 3 & 4 \\ 2 & 1 & 4 & 3 \end{pmatrix}, \quad \begin{pmatrix} 1 & 2 & 3 & 4 \\ 3 & 2 & 1 & 4 \end{pmatrix},$$

$$\begin{pmatrix} 1 & 2 & 3 & 4 \\ 4 & 3 & 2 & 1 \end{pmatrix}, \quad \begin{pmatrix} 1 & 2 & 3 & 4 \\ 1 & 4 & 3 & 2 \end{pmatrix}.$$

Given a permutation $\sigma \in P_n$, it follows by 5.1(1) that

$$\sigma^2 = \sigma \circ \sigma \in P_n.$$

More generally, for every positive integer k we have $\sigma^k \in P_n$ where

$$\sigma^k = \underbrace{\sigma \circ \sigma \circ \cdots \circ \sigma}_{k}.$$

Similarly, we have $\sigma^{-1} \in P_n$ and $\sigma^{-k} = (\sigma^{-1})^k \in P_n$.

Example Consider the symmetries of the square. Let

$$\sigma = \begin{pmatrix} 1 & 2 & 3 & 4 \\ 2 & 3 & 4 & 1 \end{pmatrix}$$

be the clockwise rotation through $90°$. Then

$$\sigma^2 = \begin{pmatrix} 1 & 2 & 3 & 4 \\ 2 & 3 & 4 & 1 \end{pmatrix} \circ \begin{pmatrix} 1 & 2 & 3 & 4 \\ 2 & 3 & 4 & 1 \end{pmatrix}$$

$$= \begin{pmatrix} 1 & 2 & 3 & 4 \\ 3 & 4 & 1 & 2 \end{pmatrix}.$$

To compute σ^{-1}, observe that

$$\sigma(1) = 2 \Longrightarrow \sigma^{-1}(2) = 1$$
$$\sigma(2) = 3 \Longrightarrow \sigma^{-1}(3) = 2$$
$$\sigma(3) = 4 \Longrightarrow \sigma^{-1}(4) = 3$$
$$\sigma(4) = 1 \Longrightarrow \sigma^{-1}(1) = 4$$

and so

$$\sigma^{-1} = \begin{pmatrix} 1 & 2 & 3 & 4 \\ 4 & 1 & 2 & 3 \end{pmatrix}.$$

We also have

$$\sigma^{-2} = \begin{pmatrix} 1 & 2 & 3 & 4 \\ 3 & 4 & 1 & 2 \end{pmatrix},$$

so that $\sigma^2 = \sigma^{-2}$. Geometrically this merely corresponds to the fact that a clockwise rotation through $180°$ is the same as an anti-clockwise rotation through $180°$.

Given $\sigma \in P_n$, consider now the relation $\sim$ defined on the set $X = \{1, \ldots, n\}$ by

$$a \sim b \iff (\exists k \in \mathbb{Z}) \ \sigma^k(a) = b.$$

5.4 Theorem $\sim$ *is an equivalence relation.*

Proof (1) For every $a \in X$ we have $\sigma^0(a) = a$, so $a \sim a$ and $\sim$ is reflexive.

(2) If $a \sim b$ then $\sigma^k(a) = b$ for some $k \in \mathbb{Z}$. It follows that $\sigma^{-k}(b) = a$ so $b \sim a$ and $\sim$ is symmetric.

(3) Suppose that $a \sim b$ and $b \sim c$. Then there exist $k, m \in \mathbb{Z}$ such that $\sigma^k(a) = b$ and $\sigma^m(b) = c$. It follows that

$$\sigma^{m+k}(a) = (\sigma^m \circ \sigma^k)(a) = \sigma^m(b) = c$$

so $a \sim c$ and $\sim$ is also transitive. $\Diamond$

Definition If σ is a permutation on X then the $\sim$-classes are called the *orbits* of σ. If $A \subseteq X$ is an orbit of σ and $|A| = k$ then we say that A is an *orbit of size k*. By a *fixed point* of σ we shall mean any element of X that is contained in an orbit of size 1.

The calculation of orbits is simplified by the following result.

5.5 Theorem *Let σ be a permutation on X. Then for every $a \in X$ the orbit of σ that contains a is*

$$A = \{\sigma(a), \sigma^2(a), \ldots, \sigma^{k-1}(a), \sigma^k(a) = a\}.$$

Proof Using the transitivity of the relation $\sim$, we see that every element of A is indeed in the orbit of a. Conversely, suppose that $b \in X$ is such that $b \sim a$. Then $\sigma^m(a) = b$ for some $m \in \mathbb{Z}$. By euclidean division, we have

$$m = kq + r \text{ for } 0 \leq r < k.$$

Consequently

$$\begin{aligned} b = \sigma^m(a) &= \sigma^{kq+r}(a) \\ &= (\sigma^r \circ \underbrace{\sigma^k \circ \cdots \circ \sigma^k}_{q})(a) \\ &= \sigma^r(a). \end{aligned}$$

But since $0 \leq r < k$ we have $\sigma^r(a) \in A$, and so $b \in A$ as required. $\Diamond$

Example Consider the permutation

$$\sigma = \begin{pmatrix} 1 & 2 & 3 & 4 & 5 & 6 & 7 & 8 \\ 4 & 5 & 3 & 6 & 2 & 8 & 7 & 1 \end{pmatrix} \in P_8.$$

To compute the orbits of σ observe that

$$\sigma(1) = 4, \quad \sigma(4) = 6, \quad \sigma(6) = 8, \quad \sigma(8) = 1,$$

and so $1 \sim 4, 4 \sim 6, 6 \sim 8, 8 \sim 1$. Thus $\{1, 4, 6, 8\}$ is an orbit. Also, $\{2, 5\}$ is an orbit, as are $\{3\}$ and $\{7\}$. The partition into orbits is therefore

$$\{\{1, 4, 6, 8\}, \{2, 5\}, \{3\}, \{7\}\}.$$

The fixed points of σ are 3 and 7.

Definition A permutation $\sigma \in P_n$ is called a *cycle* if σ has at most one orbit of size greater than 1. If every orbit is of size 1 then σ is the identity permutation id_X which we call a *1-cycle*. If $\sigma \neq \mathrm{id}_X$ is a cycle with an orbit of size $r > 1$ then we call σ an *r-cycle*.

Example The permutation $\sigma \in P_6$ given by

$$\sigma = \begin{pmatrix} 1 & 2 & 3 & 4 & 5 & 6 \\ 3 & 5 & 2 & 4 & 1 & 6 \end{pmatrix}$$

is a 4-cycle. The orbits of σ are

$$\{1, 3, 2, 5\}, \quad \{4\}, \quad \{6\}.$$

The permutation $\tau \in P_6$ given by

$$\tau = \begin{pmatrix} 1 & 2 & 3 & 4 & 5 & 6 \\ 3 & 5 & 2 & 6 & 4 & 1 \end{pmatrix}$$

is a 6-cycle, for τ has the single orbit

$$\{1, 3, 2, 5, 4, 6\}.$$

Example P_n contains $(n-1)!$ permutations which are n-cycles. To see this, suppose that $\sigma : X \to X$ is an n-cycle. Then since $\sigma(1) \neq 1$ there are $n-1$ choices of $\sigma(1)$. Next,

$$\sigma^2(1) \in X \setminus \{1, \sigma(1)\}$$

and so there are $n-2$ choices of $\sigma^2(1)$. Continuing in this way, we require

$$\sigma^i(1) \in X \setminus \{1, \sigma(1), \sigma^2(1), \ldots, \sigma^{i-1}(1)\}$$

and so there are $n-i$ choices of $\sigma^i(1)$. Hence there are

$$(n-1)(n-2) \cdots 2 \cdot 1 = (n-1)!$$

choices of σ so that σ is an n-cycle.

The notation that we have introduced for permutations does not emphasise the special properties of cycles. For many purposes, the notation that we shall now introduce is superior.

Let $\sigma \in P_n$ be a k-cycle. If $k > 1$ then by 5.5 there is an orbit

$$\{\sigma(a) \, \sigma^2(a) \, \ldots \, \sigma^{k-1}(a) \, \sigma^k(a) = a\},$$

and all other points are fixed by σ. We shall write, in *cycle notation*,

$$\sigma = \big(a, \sigma(a), \sigma^2(a), \ldots, \sigma^{k-1}(a)\big).$$

In the case where $k = 1$ we have the identity permutation id which we shall write as (1).

Example The permutations $\sigma, \tau \in P_6$ given by

$$\sigma = \begin{pmatrix} 1 & 2 & 3 & 4 & 5 & 6 \\ 3 & 5 & 2 & 4 & 1 & 6 \end{pmatrix}, \quad \tau = \begin{pmatrix} 1 & 2 & 3 & 4 & 5 & 6 \\ 3 & 5 & 2 & 6 & 4 & 1 \end{pmatrix}$$

are, in cycle notation, expressed as

$$\sigma = (1\ 3\ 2\ 5), \quad \tau = (1\ 3\ 2\ 5\ 4\ 6).$$

Our objective now is to show how an arbitrary permutation can be decomposed into a composite of cycles, such cycles being chosen in a rather special way.

Definition If $\sigma \in P_n$ then we say that $x \in X$ is *moved by* σ if x is not a fixed point of σ. If $\sigma, \tau \in P_n$ and no point of X is moved by both σ and τ then we say that σ and τ are *disjoint*.

Example The permutations $\sigma, \tau \in P_8$ given by

$$\sigma = \begin{pmatrix} 1 & 2 & 3 & 4 & 5 & 6 & 7 & 8 \\ 6 & 7 & 3 & 1 & 5 & 4 & 2 & 8 \end{pmatrix}$$

$$\tau = \begin{pmatrix} 1 & 2 & 3 & 4 & 5 & 6 & 7 & 8 \\ 1 & 2 & 8 & 4 & 3 & 6 & 7 & 5 \end{pmatrix}$$

are disjoint.

5.6 Theorem *If $\sigma, \tau \in P_n$ are disjoint then they commute, in the sense that $\sigma \circ \tau = \tau \circ \sigma$.*

Proof Suppose that $a \in X$ is moved by σ. Then, since σ and τ are disjoint, a must be a fixed point of τ. Observe that a and $\sigma(a)$ belong to the same orbit of σ, so $\sigma(a)$ is also moved by σ and fixed by τ. Thus we have

$$(\tau \circ \sigma)(a) = \sigma(a) = (\sigma \circ \tau)(a).$$

A similar argument shows that $(\tau \circ \sigma)(a) = (\sigma \circ \tau)(a)$ for every $a \in X$ that is fixed by σ. We conclude that σ and τ commute. ◇

5.7 Theorem *Every $\sigma \in P_n$ can be written as a composite of disjoint cycles. Moreover, omitting 1-cycles, this decomposition is unique up to the order in which the disjoint cycles occur.*

Proof Suppose that the orbits of σ are $A_1, \ldots, A_t$. Then clearly $\{A_1, \ldots, A_t\}$ is a partition of X. Define $\tau_1, \ldots, \tau_t \in P_n$ by setting

$$\tau_i(a) = \begin{cases} \sigma(a) & \text{if } a \in A_i; \\ a & \text{if } a \notin A_i. \end{cases}$$

Then it is readily seen that each τ_i is a cycle. Observe that if $|A_i| = 1$ then $\tau_i = \mathrm{id}_X$. Now it is clear from this definition that the τ_i are disjoint. Also,

$$\sigma = \tau_1 \circ \tau_2 \circ \cdots \circ \tau_t.$$

In fact, given $a \in X$, there is a unique set A_i with $a \in A_i$. Then $\tau_j(a) = a$ for $j \neq i$ and so, since $\tau_1, \ldots, \tau_t$ commute, we have

$$(\tau_1 \circ \tau_2 \circ \cdots \circ \tau_t)(a) = \tau_i(a) = \sigma(a).$$

Omitting 1-cycles from the decomposition, we see that we can write

$$\sigma = \tau_1 \circ \tau_2 \circ \cdots \circ \tau_p$$

where the τ_i are disjoint r_i-cycles for $r_i \geq 2$. Suppose that we also have

$$\sigma = \rho_1 \circ \rho_2 \circ \cdots \circ \rho_q$$

where the ρ_i are disjoint s_i-cycles for $s_i \geq 2$. If $a \in X$ is moved by the cycle τ_1 then a must also be moved by ρ_i for some i. Since the cycles ρ_j commute, we can assume without loss of generality that a is moved by ρ_1. Then a is fixed by $\rho_2, \ldots, \rho_q$. By 5.5, the orbit of a under σ is

$$\{\sigma(a), \sigma^2(a), \ldots, \sigma^{k-1}(a), \sigma^k(a) = a\},$$

and therefore $\rho_1 = \tau_1 = (a \; \sigma(a) \; \ldots \; \sigma^{k-1}(a))$. We can continue this process to pair off the τ_i with the ρ_j and thereby establish the required uniqueness. $\diamondsuit$

Observe from the proof of 5.7 that in order to find the disjoint cycle decomposition of a permutation σ we have to find the orbits of σ.

Example Consider $\sigma \in P_{10}$ given by

$$\sigma = \begin{pmatrix} 1 & 2 & 3 & 4 & 5 & 6 & 7 & 8 & 9 & 10 \\ 6 & 9 & 7 & 2 & 10 & 1 & 5 & 8 & 4 & 3 \end{pmatrix}.$$

It is readily seen that the orbit of 1 is $\{1, 6\}$, that of 2 is $\{2, 9, 4\}$, that of 3 is $\{3, 7, 5, 10\}$, and that of 8 is $\{8\}$. We therefore have

$$\sigma = (1 \; 6) \circ (2 \; 9 \; 4) \circ (3 \; 7 \; 5 \; 10).$$

When a permutation is given as a composite of cycles that are not necessarily disjoint, it is a straightforward matter to express it as a composite of cycles that are disjoint.

Example Consider $\sigma \in P_6$ given by

$$\sigma = (1\ 2\ 5) \circ (3\ 1\ 6) \circ (4\ 2) \circ (5\ 6\ 2).$$

We first compute the orbit of 1 by applying each cycle in turn, starting of course with the rightmost one since we are dealing with composites of mappings. The orbit of 1 is

$$\{1, 6, 4, 5, 3, 2\},$$

and hence we see that $\sigma = (1\ 6\ 4\ 5\ 3\ 2)$.

Example Let $\tau \in P_6$ be given by

$$\tau = (1\ 2\ 3) \circ (2\ 3\ 4) \circ (3\ 4\ 5) \circ (4\ 5\ 6).$$

The orbit of 1 is $\{1, 2\}$, that of 3 is $\{3\}$, that of 4 is $\{4\}$, and that of 5 is $\{5, 6\}$. We therefore have $\tau = (1\ 2) \circ (5\ 6)$.

Example The permutation $\lambda_5 : \mathbb{Z}_{11}^\star \to \mathbb{Z}_{11}^\star$ considered in the Example immediately following 5.3 can be written as

$$(1\ 5\ 3\ 4\ 9) \circ (2\ 10\ 6\ 8\ 7).$$

The permutation λ_{10} can be written as

$$\lambda_{10} = (1\ 10) \circ (2\ 9) \circ (3\ 8) \circ (4\ 7) \circ (5\ 6).$$

Given a permutation $\sigma \in P_n$ we have $\sigma^i \in P_n$ for every $i \in \mathbb{Z}$. Since $|P_n| = n!$, not all the powers σ^i can be distinct. We must therefore have $\sigma^i = \sigma^j$ for some i, j with $i \neq j$. Composing with σ^{-j}, we obtain $\sigma^{i-j} = \text{id}$. Now by the well ordering property of $\mathbb{N}$ there must exist a least positive integer k such that $\sigma^k = \text{id}$.

Definition For each $\sigma \in P_n$ the least positive integer k such that $\sigma^k = \text{id}$ is called the *order* of σ.

5.8 Theorem *Let* $\sigma \in P_n$ *be such that*

$$\sigma = \rho \circ \tau$$

where ρ *and* τ *are disjoint. Then the order of* σ *is the lowest common multiple of the orders of* ρ *and* τ.

Proof Let σ have order m and let ρ, τ have orders r, t respectively. By 5.6, ρ and τ commute and so, for every $k \in \mathbb{Z}$,

$$\sigma^k = (\rho \circ \tau)^k = \rho^k \circ \tau^k.$$

We then have

$$\mathrm{id} = \sigma^m = \rho^m \circ \tau^m.$$

Since ρ and τ are disjoint, so also are ρ^m and τ^m and so from $\rho^m = \tau^{-m}$ we deduce that

$$\rho^m = \mathrm{id} = \tau^m,$$

and hence that $r|m, t|m$. But if k is a positive integer such that $r|k$ and $t|k$ then
$$\sigma^k = \rho^k \circ \tau^k = \mathrm{id},$$

from which it follows that $m|k$. Thus $m = \mathrm{lcm}\{r, t\}$ and the result follows. $\diamond$

5.9 Corollary *The order of $\sigma \in P_n$ is the lowest common multiple of the lengths of the cycles in the decomposition of σ as a composite of disjoint cycles.*

Proof Using the fact that an r-cycle τ has order r, the result follows by using induction on the number of disjoint cycles. The inductive step is a consequence of 5.8. $\diamond$

Example To find the order of $\sigma \in P_{10}$ given by

$$\sigma = \begin{pmatrix} 1 & 2 & 3 & 4 & 5 & 6 & 7 & 8 & 9 & 10 \\ 6 & 9 & 7 & 2 & 10 & 1 & 5 & 8 & 4 & 3 \end{pmatrix},$$

write σ as a composite of disjoint cycles :

$$\sigma = (1\ 6) \circ (2\ 9\ 4) \circ (3\ 7\ 5\ 10).$$

The order of σ is then $\mathrm{lcm}\{2, 3, 4\} = 12$.

Example Given $\sigma \in P_n$, define a mapping $\lambda_\sigma : P_n \to P_n$ by

$$\lambda_\sigma(\rho) = \sigma \circ \rho.$$

Note that λ_σ is a bijection and so is a permutation on P_n.

Consider the case where $n = 3$. We have

$$P_3 = \{(1), (1\ 2), (1\ 3), (2\ 3), (1\ 2\ 3), (1\ 3\ 2)\}.$$

Write these six elements as follows :

$$1 = (1), \quad 2 = (1\ 2), \quad 3 = (1\ 3),$$
$$4 = (2\ 3), \quad 5 = (1\ 2\ 3), \quad 6 = (1\ 3\ 2).$$

Consider now the permutation

$$\lambda_{(1\ 2)} : P_3 \to P_3$$

defined as above. We have

$$\lambda_{(1\ 2)} = (1\ 2) \circ (3\ 6) \circ (4\ 5).$$

That $\lambda_{(1\ 2)}$ is a composite of disjoint 2-cycles is not surprising. For, given any $\rho \in P_3$ we have

$$\lambda_{(1\ 2)}^2(\rho) = (1\ 2) \circ (1\ 2) \circ \rho = \rho,$$

so $\lambda_{(1\ 2)}$ is a permutation of order 2 and so, by 5.9, is the composite of two disjoint 2-cycles. Similarly, 5.9 shows that

$$\lambda_{(1\ 2\ 3)} : P_3 \to P_3$$

is a composite of 3-cycles since, for every $\rho \in P_3$,

$$\lambda_{(1\ 2\ 3)}^3(\rho) = (1\ 2\ 3) \circ (1\ 2\ 3) \circ (1\ 2\ 3) \circ \rho = \rho.$$

In fact, it is easy to compute that

$$\lambda_{(1\ 2\ 3)} = (1\ 5\ 6) \circ (2\ 3\ 4).$$

Since $(1\ 2\ 3)^2 = (1\ 2\ 3) \circ (1\ 2\ 3) = (1\ 3\ 2)$, we see that $\lambda_{(1\ 2\ 3)}^2 = \lambda_{(1\ 3\ 2)}$ and hence that

$$\lambda_{(1\ 3\ 2)} = (1\ 5\ 6)^2 \circ (2\ 3\ 4)^2$$
$$= (1\ 6\ 5) \circ (2\ 4\ 3).$$

We can relate orders of permutations to the following idea of a partition of an integer.

Definition By a *partition* of a positive integer n we shall mean a set $\{n_1, n_2, \ldots, n_k\}$ of positive integers such that

$$n_1 + n_2 + \cdots + n_k = n.$$

Let Σ_n denote the set of all partitions of the positive integer n. We can define a mapping $p : P_n \to \Sigma_n$ as follows. Given $\sigma \in P_n$, write

$$\sigma = \tau_1 \circ \tau_2 \circ \cdots \tau_t$$

where τ_i is an r_i-cycle with $r_i > 1$ and the cycles $\tau_1, \ldots, \tau_t$ are disjoint. Let $\tau_{t+1}, \ldots, \tau_k$ be 1-cycles corresponding to the fixed points of σ, so that for $t + 1 \leq i \leq k$ each τ_i is an r_i-cycle with $r_i = 1$. We define

$$p(\sigma) = \{r_1, r_2, \ldots, r_k\}.$$

Note that the uniqueness of the disjoint cycle decomposition established in 5.7 ensures that p is indeed a mapping.

Example Taking $\sigma \in P_{10}$ as in the previous Example, we have

$$\sigma = (1\ 6) \circ (2\ 9\ 4) \circ (3\ 7\ 5\ 10),$$

so 8 is a fixed point of σ. The corresponding partition of 10 is

$$\{1, 2, 3, 4\}.$$

Note that the mapping $p : P_n \to \Sigma_n$ is always a surjection, but for $n \geq 3$ it is not an injection. For example, we have

$$(1\ 2) \overset{p}{\longmapsto} \{2, \underbrace{1, 1, \ldots, 1}_{n-2}\},$$

$$(1\ 3) \overset{p}{\longmapsto} \{2, \underbrace{1, 1, \ldots, 1}_{n-2}\}.$$

If we consider a fixed partition $A \in \Sigma_n$ then by 5.9 we see that each permutation in the set

$$\{\sigma \in P_n\ ;\ p(\sigma) = A\}$$

has the same order. The equivalence relation determined by the mapping p therefore partitions P_n into equivalence classes with the property that permutations in the same class have the same order.

Example The partition of P_3 determined by the mapping p is

$$\{\{(1)\}, \{(1\ 2), (1\ 3), (2\ 3)\}, \{(1\ 2\ 3), (1\ 3\ 2)\}\}.$$

It is possible for elements in distinct classes in the partition of P_n determined by the mapping p to have the same order.

Example In P_4, $(1\ 2)$ and $(1\ 2) \circ (3\ 4)$ have order 2. But

$$(1\ 2) \overset{p}{\longmapsto} \{2, 1, 1\},$$
$$(1\ 2) \circ (3\ 4) \overset{p}{\longmapsto} \{2, 2\},$$

so $p(1\ 2) \neq p\big((1\ 2) \circ (3\ 4)\big)$.

We can decompose a permutation into a composite of 2-cycles which may not be disjoint. In order to prove this, we consider the following type of permutation.

Definition A 2-cycle is called a *transposition*.

5.10 Theorem *Every r-cycle can be expressed as a composite of $r - 1$ transpositions.*

Proof Let $\sigma \in P_n$ be the r-cycle

$$\sigma = (i_1\ i_2\ \ldots\ i_r).$$

We show that $\sigma = \tau$ where

$$\tau = (i_1\ i_r) \circ (i_1\ i_{r-1}) \circ \cdots \circ (i_1\ i_2).$$

Let $a \in X$ and observe that if a is fixed by σ then a is also fixed by τ. Suppose then that $a = i_j$ is moved by σ. If $j \neq r$ then we have

$$\sigma(i_j) = i_{j+1}, \quad \tau(i_j) = i_{j+1}.$$

Since $\sigma(i_r) = i_1$ and $\tau(i_r) = i_1$, it follows that $\sigma = \tau$ as required. $\diamond$

5.11 Corollary *Every $\sigma \in P_n$ can be written as a composite of transpositions.*

Proof By 5.7, σ can be written as a composite of disjoint cycles. Now apply 5.10 to each cycle. $\diamond$

Example Consider $\sigma \in P_{10}$ given by

$$\sigma = \begin{pmatrix} 1 & 2 & 3 & 4 & 5 & 6 & 7 & 8 & 9 & 10 \\ 6 & 9 & 7 & 2 & 10 & 1 & 5 & 8 & 4 & 3 \end{pmatrix}.$$

Following the method of 5.11, we have

$$\sigma = (1\ 6) \circ (2\ 9\ 4) \circ (3\ 7\ 5\ 10)$$
$$= (1\ 6) \circ (2\ 4) \circ (2\ 9) \circ (3\ 10) \circ (3\ 5) \circ (3\ 7).$$

It should be observed that this decomposition into transpositions is not unique. For example, we also have

$$\sigma = (1\ 6) \circ (4\ 9) \circ (2\ 4) \circ (5\ 7) \circ (3\ 5) \circ (5\ 10).$$

The fact that these two decompositions of σ into transpositions contain the same number of transpositions is a consequence of the way in which the examples were constructed. In general, decompositions into transpositions can contain different numbers of transpositions, as we shall now show.

Example The permutation

$$\sigma = \begin{pmatrix} 1 & 2 & 3 & 4 & 5 & 6 \\ 2 & 1 & 3 & 4 & 6 & 5 \end{pmatrix}$$

can be written as a composite of transpositions in, for example, the following ways :

$$(1\ 3) \circ (1\ 2) \circ (2\ 4) \circ (2\ 3) \circ (3\ 5) \circ (3\ 4) \circ (4\ 6) \circ (4\ 5),$$
$$(1\ 3) \circ (1\ 2) \circ (4\ 5) \circ (2\ 3) \circ (4\ 6) \circ (4\ 5),$$
$$(1\ 3) \circ (1\ 2) \circ (2\ 3) \circ (5\ 6),$$
$$(1\ 2) \circ (5\ 6).$$

Observe that σ has been expressed as a composite of 2, 4, 6, and 8 transpositions. It is no coincidence that all of these numbers are even, as we now aim to show. First, we prove a preliminary result.

5.12 Theorem *Suppose that $\sigma \in P_n$ contains r cycles in its decomposition as a composite of disjoint cycles, where the fixed points of σ are included as 1-cycles. Then if τ is a transposition the permutation $\tau \circ \sigma$ contains $r - 1$ or $r + 1$ cycles in its decomposition into disjoint cycles, again counting fixed points as 1-cycles.*

Proof Suppose that $\tau = (a\,b)$ and that a, b both occur in the same disjoint cycle of σ. Since disjoint cycles commute, we can write

$$\sigma = \sigma_1 \circ \sigma_2 \circ \cdots \circ \sigma_r$$

where $\sigma_1, \ldots, \sigma_r$ are disjoint and

$$\sigma_1 = (a\,i_1\,\ldots\,i_k\,b\,i_{k+1}\,\ldots\,i_t).$$

Now the permutation

$$\tau \circ \sigma = (a\,i_1\,\ldots\,i_k) \circ (b\,i_{k+1}\,\ldots\,i_t) \circ \sigma_2 \circ \cdots \circ \sigma_r$$

and contains $r + 1$ disjoint cycles.

The other case to consider is when $\tau = (a\,b)$ and a, b occur in distinct cycles in the disjoint cycle decomposition of σ. Again using the fact that disjoint cycles commute, we can write

$$\sigma = \sigma_1 \circ \sigma_2 \circ \cdots \circ \sigma_r$$

where the cycles

$$\sigma_1 = (a\,i_1\,\ldots\,i_k), \quad \sigma_2 = (b\,i_{k+1}\,\ldots\,i_t), \quad \sigma_3, \ldots, \sigma_r$$

are disjoint. In this case,

$$\tau \circ \sigma = (a\,i_1\,\ldots\,i_k\,b\,i_{k+1}\,\ldots\,i_t) \circ \sigma_3 \circ \cdots \circ \sigma_r$$

which contains $r - 1$ disjoint cycles. $\diamond$

Example The permutation $\sigma = (1\,2) \circ (4\,5\,6) \in P_6$ has three cycles in its decomposition since we include a 1-cycle for the fixed point 3. Taking $\tau_1 = (4\,5)$, we have

$$\tau_1 \circ \sigma = (4\,5) \circ (1\,2) \circ (4\,5\,6) = (1\,2) \circ (5\,6),$$

so $\tau_1 \circ \sigma$ contains four disjoint cycles, namely

$$(1\,2), \quad (5\,6), \quad (3), \quad (4).$$

Also, if $\tau_2 = (1\,3)$ we have

$$\tau_2 \circ \sigma = (1\,3) \circ (1\,2) \circ (4\,5\,6) = (1\,2\,3) \circ (4\,5\,6),$$

so $\tau_2 \circ \sigma$ contains two disjoint cycles, namely $(1\,2\,3)$ and $(4\,5\,6)$.

5.13 Theorem *Given* $\sigma \in P_n$, *suppose that*

$$\sigma = \tau_1 \circ \tau_2 \circ \cdots \circ \tau_s,$$
$$\sigma = \tau_1' \circ \tau_2' \circ \cdots \circ \tau_{s'}'$$

are two expressions for σ *as composites of transpositions. Then* s *and* s' *have the same parity, in the sense that either they are both even or they are both odd.*

Proof The transposition τ_s has one 2-cycle and $n - 2$ fixed points so, according to 5.12, τ_s has $n - 1$ cycles. Then

$$\tau_{s-1} \circ \tau_s$$

contains $(n - 1) + 1$ or $(n - 1) - 1$ cycles. We can continue in this way to apply 5.12 to compute the number of cycles in

$$\tau_{s-2} \circ \tau_{s-1} \circ \tau_s,$$

and so on. Suppose that in the $s - 1$ applications of 5.12 a cycle is added a times, so that a cycle is subtracted $s - 1 - a$ times. Then σ contains

$$(n - 1) + a - (s - 1 + a) = n - s + 2a$$

cycles. Using precisely the same argument with the second expression for σ and assuming that a cycle is added b times and subtracted $s' - 1 - b$ times, we compute the number of cycles in σ to be

$$n - s' + 2b.$$

We thus have $n - s + 2a = n - s' + 2b$ whence $s - s' = 2(a - b)$ and so $s - s'$ is even. It follows that s, s' are either both even or both odd. $\Diamond$

Definition A permutation is said to be *even* if it can be expressed as a composite of an even number of transpositions, and *odd* if it can be expressed as a composite of an odd number of transpositions.

5.14 Theorem *Let* $\sigma \in P_n$ *and suppose that* σ *contains* r *cycles in its decomposition as a composite of disjoint cycles, 1-cycles being counted for the fixed points. Then* σ *is an even permutation if* $n - r$ *is even, and an odd permutation if* $n - r$ *is odd.*

Proof Suppose that in the decomposition of σ the r cycles have lengths $t_1, \ldots, t_r$. Since each point appears in precisely one cycle, we have

$$\sum_{i=1}^{r} t_i = n.$$

But, using 5.10, we see that a cycle of length t_i can be written as a composite of $t_i - 1$ transpositions. Hence σ can be written as a composite of

$$\sum_{i=1}^{r} (t_i - 1) = \sum_{i=1}^{r} t_i - r = n - r$$

transpositions. The result now follows. $\diamond$

Example The permutation $\sigma = (1\ 2\ 3) \circ (4\ 5) \in P_6$ is odd. For, σ has three cycles, namely $(1\ 2\ 3)$, $(4\ 5)$ and (6); and $6 - 3 = 3$ which is odd. It is easy to see that $\sigma = (1\ 3) \circ (1\ 2) \circ (4\ 5)$.

Example The 'Fifteen Puzzle', invented by Sam Loyd and extremely popular for several years after its appearance in 1878, consists of fifteen blocks numbered from 1 to 15 contained within a frame. A move consists of sliding a block into the one empty space. The starting position is the following :

1	2	3	4
5	6	7	8
9	10	11	12
13	14	15	

Observe that any sequence of moves that makes the empty space return to the bottom right-hand corner of the frame is an even permutation. For, denoting the empty space by E, we see that moving block X into the empty space E is described by the transposition $(X\ E)$. If E is to return to the bottom right-hand corner then the number of moves left must equal the number of moves right, and the number of moves up must equal the number of moves down. Therefore the resulting permutation contains an even number of transpositions.

Is, for example, the position shown below possible?

10	9	7	4
8	3	6	13
5	2	12	15
1	11	14	

Consider the permutation $\sigma \in P_{16}$ of the fifteen blocks and one space E that is required to transform the starting position to this final position. We see that

$$\sigma = (1\ 10\ 2\ 9\ 5\ 8\ 13) \circ (3\ 7\ 6) \circ (11\ 12\ 15\ 14)$$

with two fixed points, namely 4 and E. By 5.14 we see that, since σ has five cycles,

$$n - r = 16 - 5 = 11,$$

so σ is an odd permutation. It is therefore impossible to reach the position indicated from the original starting position.

5.15 Theorem *For $n \geq 2$ the number of even permutations in P_n is $\frac{1}{2}n!$.*

Proof If $\tau = (1\ 2)$ then the mapping $f_\tau : P_n \to P_n$ given by

$$\rho \mapsto f_\tau(\rho) = \tau \circ \rho$$

is a bijection. Observe that f_τ maps even permutations to odd permutations, so if A is the subset of even permutations in P_n then $f_\tau^{\rightarrow}(A) = \{f_\tau(a)\ ;\ a \in A\}$ is the set of odd permutations. Now

$$P_n = A \cup f_\tau^{\rightarrow}(A)$$

and clearly

$$A \cap f_\tau^{\rightarrow}(A) = \emptyset.$$

Since f_τ is a bijection we have $|A| = |f_\tau^{\rightarrow}(A)|$ so

$$2|A| = |P_n| = n!,$$

from which the result follows. $\diamondsuit$

Definition For every $\sigma \in P_n$ the *signum* of σ is defined to be

$$\epsilon_\sigma = \prod_{i<j} \frac{\sigma(j) - \sigma(i)}{j - i}.$$

Example If $\sigma = (1\ 2)$ then

$$\epsilon_\sigma = \frac{\sigma(2) - \sigma(1)}{2 - 1} = \frac{1 - 2}{2 - 1} = -1.$$

Example If $\sigma = (1\ 3\ 2)$ then

$$\begin{aligned}
\epsilon_\sigma &= \frac{\sigma(2) - \sigma(1)}{2 - 1} \cdot \frac{\sigma(3) - \sigma(1)}{3 - 1} \cdot \frac{\sigma(3) - \sigma(2)}{3 - 2} \\
&= \frac{1 - 3}{2 - 1} \cdot \frac{2 - 3}{3 - 1} \cdot \frac{2 - 1}{3 - 2} \\
&= 1.
\end{aligned}$$

5.16 Theorem *For every $\sigma \in P_n$,*

$$\epsilon_\sigma = \begin{cases} 1 & \text{if } \sigma \text{ is even;} \\ -1 & \text{if } \sigma \text{ is odd.} \end{cases}$$

Moreover,

$$(\forall \sigma, \rho \in P_n) \qquad \epsilon_{\sigma \circ \rho} = \epsilon_\sigma \epsilon_\rho.$$

Proof We show first that if τ is a transposition then $\epsilon_\tau = -1$. Suppose that τ interchanges a and b with $a < b$. Consider the product

$$\prod_{i<j} \big(\tau(j) - \tau(i)\big).$$

This has the following factors :

(1) $j - i$ where $i, j \notin \{a, b\}$;
(2) $b - i$ where $i \notin \{a, b\}, j = a$;
(3) $a - i$ where $i \notin \{a, b\}, j = b$;
(4) $a - b$;
(5) $j - b$ where $i = a, j \notin \{a, b\}$;
(6) $j - a$ where $i = b, j \notin \{a, b\}$.

Now the expressions in (1), (2) and (6) are all positive, and that in (4) is negative. As for (3), a negative term occurs when $a < i < b$; and in (5) a negative term occurs when $a < j < b$. The total number of negative terms is therefore odd, and it follows that

$$\prod_{i<j} \big(\tau(j) - \tau(i)\big) = -\prod_{i<j} (j - i)$$

and hence that $\epsilon_\tau = -1$.

It is now immediate that, for every $\sigma \in P_n$,

$$\epsilon_\sigma = \left\{ \begin{array}{ll} 1 & \text{if } \sigma \text{ is even;} \\ -1 & \text{if } \sigma \text{ is odd.} \end{array} \right.$$

Consideration of how even and odd permutations compose now shows that

$$\begin{aligned} \epsilon_{\sigma \circ \rho} &= \left\{ \begin{array}{ll} 1 & \text{if } \sigma, \rho \text{ are both even or both odd;} \\ -1 & \text{otherwise,} \end{array} \right. \\ &= \epsilon_\sigma \epsilon_\rho. \quad \Diamond \end{aligned}$$

5.17 Corollary $\epsilon_{\sigma^{-1}} = \epsilon_\sigma$.

Proof σ^{-1} is even or odd according to whether σ is even or odd. $\Diamond$

Example If $\sigma \in P_8$ is given by

$$\sigma = \begin{pmatrix} 1 & 2 & 3 & 4 & 5 & 6 & 7 & 8 \\ 3 & 7 & 2 & 6 & 5 & 4 & 8 & 1 \end{pmatrix}$$

then we have

$$\begin{aligned} \sigma &= (1\ 3\ 2\ 7\ 8) \circ (4\ 6) \\ &= (1\ 8) \circ (1\ 7) \circ (1\ 2) \circ (1\ 3) \circ (4\ 6), \end{aligned}$$

so σ is odd and $\epsilon_\sigma = -1$.

The signum of a permutation is particularly important in the evaluation of determinants of square matrices (see, for example, Volume Two).

CHAPTER SIX

Cardinals and the natural numbers

It can be justifiably argued that during the last century the theory of sets has revolutionized the whole of mathematics. One of the major achievements that has resulted from the development of set theory is a clear interpretation of the concept of 'infinity'. Our objective in this final Chapter is to make this concept precise, and in so doing give a concrete interpretation of what a natural number really is. The reader will appreciate as we proceed that all of what follows rests solely on the concepts of sets and mappings.

We begin with the hypothesis, due to Cantor, that associated with every set E there is a set, called the *cardinal* of E and written $|E|$, such that

(1) $|E|$ is equipotent to E;
(2) F is equipotent to E if and only if $|F| = |E|$.

Roughly speaking, $|E|$ is an elected representative from the class of E relative to the equivalence relation 'is equipotent to'.

In what follows we shall find it convenient to describe arbitrary cardinals by bold faced letters, such as $\mathbf{a}, \mathbf{b}, \mathbf{c}$. Note from (1) and (2) above that for every cardinal $\mathbf{a}$ we have $|\mathbf{a}| = \mathbf{a}$. For, $|\mathbf{a}|$ is the representative in the class of $\mathbf{a}$; and by definition this is $\mathbf{a}$.

6.1 Theorem *If S is a set of cardinals then the relation $\leq$ defined on S by*

$$\mathbf{a} \leq \mathbf{b} \iff \text{there is an injection of } \mathbf{a} \text{ into } \mathbf{b}$$

is an order.

Proof It is clear that $\leq$ is reflexive on S. Since a composite of two injections is an injection, $\leq$ is also transitive. That it is also anti-symmetric, and hence an order, follows by the Schröder-Bernstein Theorem; for if $a \leq b$ and $b \leq a$ then there is an injection $f : a \to b$ and an injection $g : b \to a$, so a and b are equipotent whence $a = |a| = |b| = b$. $\diamond$

More can be said about this order. It can in fact be shown that the following result holds.

6.2 Theorem [Zermelo] *Every set of cardinals is well ordered and hence totally ordered under the ordering* $\leq$. $\diamond$

The well ordering implies that given cardinals a, b we have either $a \leq b$ or $b \leq a$; for, the set $S = \{a, b\}$, being well ordered, has a smallest element. Hence the total ordering is a consequence of the well ordering. Equivalently, given sets A, B there is either an injection from A to B, or an injection from B to A. In order to prove such a general result, it turns out that we have to be more precise about what we mean by a 'set'. As such a development would take us too far into the foundations of set theory, and certainly well beyond the present 'naive' level, we shall accept 6.2 without proof.

Two important cardinals that we shall require are $0 = |\emptyset|$ and $1 = |\{\emptyset\}| = |\{x\}|$ for any object x. We can extend the courtesy of regarding $\emptyset$ as a subset of every set to include the statement $0 < 1$.

For reasons that will become clear in due course, we shall now show how to define an addition and a multiplication of cardinals. For this purpose, recall that cardinals are sets, so we can form their cartesian product.

Definition If a, b are cardinals then we define the *product* ab to be the cardinal

$$ab = |a \times b|.$$

6.3 Theorem *If a, b, c are cardinals then*

(1) $ab = ba$;
(2) $a(bc) = (ab)c$;
(3) $1a = a$;
(4) $0a = 0$.

Proof (1) follows from the fact that $a \times b$ and $b \times a$ are equipotent under the bijection $(x, y) \mapsto (y, x)$; (2) follows from the fact that $a \times (b \times c)$ and $(a \times b) \times c$ are equipotent under the bijection $(x, (y, z)) \mapsto ((x, y), z)$; (3) follows from the fact that $\{\emptyset\} \times a$ and a are equipotent under the bijection $(\emptyset, x) \mapsto x$; and (4) follows from the fact that $\emptyset \times a = \emptyset$. $\diamond$

Definition If a, b are cardinals let A, B be *disjoint* sets such that $a = |A|$ and $b = |B|$. Then we define the *sum* $a + b$ to be the cardinal

$$a + b = |A \cup B|.$$

Note that for any cardinals a, b there is no difficulty in finding *disjoint* sets A, B with $a = |A|$ and $b = |B|$. For example, taking

$$A = a \times \{0\}, \quad B = b \times \{1\}$$

we have $A \cap B = \emptyset$ and, by 6.3, $|A| = a1 = a$ and $|B| = b1 = b$.

6.4 Theorem *If* a, b, c *are cardinals then*

(1) $a + b = b + a$;

(2) $a + (b + c) = (a + b) + c$;

(3) $a + 0 = a$.

Proof (1) This is immediate from the fact that $A \cup B = B \cup A$.

(2) Let A, B, C be sets with $|A| = a, |B| = b, |C| = c$ and such that $A \cap B = B \cap C = C \cap A = \emptyset$. That such a choice is possible is readily seen by defining $2 = |\{\emptyset, \{\emptyset\}\}|$ and taking

$$A = a \times \{0\}, \quad B = b \times \{1\}, \quad C = c \times \{2\}.$$

Since A, B, C are pairwise disjoint we have

$$A \cap (B \cup C) = (A \cap B) \cup (A \cap C) = \emptyset \cup \emptyset = \emptyset;$$
$$(A \cup B) \cap C = (A \cap C) \cup (B \cap C) = \emptyset \cup \emptyset = \emptyset,$$

and hence

$$a + (b + c) = |A| + |B \cup C|$$
$$= |A \cup B \cup C|$$
$$= |A \cup B| + |C|$$
$$= (a + b) + c.$$

(3) This is clear since $A \cup \emptyset = A$ and $A \cap \emptyset = \emptyset$. $\diamond$

Multiplication of cardinals is distributive over addition :

6.5 Theorem *If* a, b, c *are cardinals then* $a(b + c) = ab + ac$.

Proof Let $|A| = a, |B| = b, |C| = c$ and choose B, C to be disjoint. Then we have

$$(A \times B) \cap (A \times C) = A \times (B \cap C) = A \times \emptyset = \emptyset$$

and hence

$$\begin{aligned} a(b + c) &= |A \times (B \cup C)| \\ &= |(A \times B) \cup (A \times C)| \\ &= |A \times B| + |A \times C| \\ &= ab + ac. \quad \diamond \end{aligned}$$

Concerning the order relation $\leq$ defined above, we have the following results.

6.6 Theorem *If* a, b, c *are cardinals then*

$$a \leq b \implies \left\{ \begin{array}{l} ac \leq bc; \\ a + c \leq b + c. \end{array} \right.$$

Proof Let $a \leq b$, so that there is an injection $f : a \to b$, and define

$$g : a \times c \to b \times c$$

by the prescription $g(a, c) = \big(f(a), c\big)$. Clearly, g is also an injection and so

$$ac = |a \times c| \leq |b \times c| = bc.$$

Suppose now that c is chosen such that $a \cap c = \emptyset = b \cap c$, such a choice always being possible. Then $h : a \cup c \to b \cup c$ given by

$$h(x) = \left\{ \begin{array}{ll} f(x) & \text{if } x \in a; \\ x & \text{if } x \in c, \end{array} \right.$$

is also an injection. Consequently

$$a + c = |a \cup c| \leq |b \cup c| = b + c. \quad \diamond$$

6.7 Theorem *If* a, b *are cardinals then* $a \leq b$ *if and only if there is a cardinal* c *such that* $b = a + c$.

Proof If $a \le b$ then there is an injection $f : a \to b$ and, by 2.6, a bijection $f^+ : a \to \operatorname{Im} f$. Thus $|\operatorname{Im} f| = a$. If now C is the complement of $\operatorname{Im} f$ in b and $c = |C|$ then

$$b = |\operatorname{Im} f| + |C| = a + c.$$

Conversely, if there is a cardinal c such that $b = a + c$ then b is the union of disjoint sets A, C with $|A| = a, |C| = c$. The restriction to A of id_b is then an injection from A into b and we have $a \le b$. $\diamondsuit$

The reader will have noticed by now that the above results give properties that are reminiscent of the basic properties of the chain $\mathbb{N}$ of natural numbers. The following results strengthen this similarity.

6.8 Theorem *If* a, b *are cardinals then*

$$a + 1 = b + 1 \Longrightarrow a = b.$$

Proof Suppose that $a + 1 = b + 1 = x$, say. Then there is a set X with $|X| = x$ which has subsets A, B and elements α, β such that

$$|A| = a, \quad |B| = b, \quad C_X(A) = \{\alpha\}, \quad C_X(B) = \{\beta\}.$$

There are two cases to consider.

(1) $\alpha = \beta$: In this case $A = B$ and so $a = b$.

(2) $\alpha \ne \beta$: In this case we have $\beta \in A$ and $\alpha \in B$ whence

$$(A \cap B) \cup \{\alpha\} = (A \cup \{\alpha\}) \cap B = X \cap B = B;$$
$$(A \cap B) \cup \{\beta\} = A \cap (B \cup \{\beta\}) = A \cap X = A.$$

Since $A \cap B \cap \{\alpha\} = \emptyset = A \cap B \cap \{\beta\}$, it follows that

$$\begin{aligned}
a = |A| &= |\{\beta\}| + |A \cap B| \\
&= 1 + |A \cap B| \\
&= |\{\alpha\}| + |A \cap B| \\
&= |B| = b. \quad \diamondsuit
\end{aligned}$$

6.9 Theorem *If* a *is a cardinal then* $a \le a + 1$.

Proof If $a = 0$ then $0 < 1 = 0 + 1$. If $a \neq 0$ let $A \neq \emptyset$ be such that $|A| = a$. Then $(A \times \{\emptyset\}) \cup \{\emptyset\}$ has cardinal $a + 1$ and the mapping

$$f : A \to (A \times \{\emptyset\}) \cup \{\emptyset\}$$

described by $f(x) = (x, \emptyset)$ is an injection, so $a \leq a + 1$. $\diamond$

We now observe that it is possible for a cardinal a to be such that $a = a + 1$.

Example If $a, b \in \mathbb{R}$ with $a < b$ then we know that the closed interval $[a, b]$ is equipotent to the half-open interval $[a, b[$. Let each have cardinal a. Then since $[a, b]$ is the disjoint union of $[a, b[$ and $\{b\}$, it follows that $a = a + 1$.

Definition A cardinal a is said to be *finite* if $a < a + 1$, and *infinite* if $a = a + 1$. A set E is said to be *finite* if $|E|$ is finite, and *infinite* if $|E|$ is infinite.

6.10 Theorem *A set E is infinite if and only if there is a subset $F \neq E$ that is equipotent to E.*

Proof Suppose first that E is infinite, so that $|E| = |E| + 1$. If α is an object that does not belong to E there is then a bijection

$$f : E \cup \{\alpha\} \to E.$$

The restriction $f_E : E \to E$ is then an injection with

$$\text{Im } f_E = E \setminus \{f(\alpha)\},$$

which is a proper subset of E and is, by 2.6, equipotent to E.

Conversely, suppose that $F \subset E$ with $|F| = |E|$. Then there exists $\alpha \in C_E(F)$ and we have

$$F \subset F \cup \{\alpha\} \subseteq E.$$

Consideration of the restrictions of id_E to F and to $F \cup \{\alpha\}$ now shows that

$$|F| \leq |F| + 1 \leq |E|.$$

Since $|F| = |E|$ by hypothesis, it follows that $|E| = |E| + 1$ and so E is infinite. $\diamond$

6.11 Corollary *A set E is finite if and only if the only subset of E that is equipotent to E is E itself.* ◇

6.12 Corollary *If E is a finite set then a mapping $f : E \to E$ is injective if and only if it is surjective.*

Proof Suppose that f is injective. Then E is equipotent to the subset $\operatorname{Im} f$. Since E is finite it follows by 6.11 that $\operatorname{Im} f = E$ and so f is also surjective.

Conversely, if f is surjective then by 2.4 there is a mapping $g : E \to E$ such that $f \circ g = \operatorname{id}_E$. By 2.3, g is injective and so, by the previous paragraph, g is also surjective. Hence g is a bijection. Since its inverse must be f, we see that f is also a bijection and hence is injective. ◇

Note that 6.12 is the same as 5.2; the proof we have just given is appropriate to the discussion in hand.

One of the fundamental axioms in the theory of sets is that there exists an infinite set. It can be shown that this is equivalent to saying that the finite cardinals form a *set*. We shall denote the set of finite cardinals by $\mathbb{F}$.

To see that $\mathbb{F}$ is infinite, consider the mapping $\vartheta : \mathbb{F} \to \mathbb{F}$ given by

$$\vartheta(\mathbf{a}) = \mathbf{a} + \mathbf{1}.$$

It follows from 6.8 that ϑ is injective. But ϑ is not surjective, since for every $\mathbf{a} \in \mathbb{F}$ we have, by 6.6,

$$\vartheta(\mathbf{a}) = \mathbf{a} + \mathbf{1} \geq \mathbf{0} + \mathbf{1} = \mathbf{1} > \mathbf{0}.$$

and so there is no $\mathbf{a} \in \mathbb{F}$ such that $\vartheta(\mathbf{a}) = \mathbf{0}$. It follows by 6.12 that $\mathbb{F}$ must be infinite.

6.13 Theorem *If $\mathbf{n}$ is a finite cardinal then every cardinal $\mathbf{a}$ with $\mathbf{a} \leq \mathbf{n}$ is also finite.*

Proof By 6.7, there is a cardinal $\mathbf{b}$ such that $\mathbf{n} = \mathbf{a} + \mathbf{b}$. Then, by 6.4,

$$(\mathbf{a} + \mathbf{1}) + \mathbf{b} = (\mathbf{a} + \mathbf{b}) + \mathbf{1} = \mathbf{n} + \mathbf{1} \neq \mathbf{n} = \mathbf{a} + \mathbf{b},$$

from which we deduce that $\mathbf{a} + \mathbf{1} \neq \mathbf{a}$ and so $\mathbf{a}$ is finite. ◇

The set $\mathbb{F}$ of finite cardinals has the following properties.

6.14 Theorem *For every cardinal* n,

(1) $0 \leq n$;

(2) *if* $0 < n \leq 1$ *then* $n = 1$.

Proof (1) By 6.4, $n = n + 0$ so (1) follows from 6.7.

(2) If $0 < n \leq 1$, let X be such that $|X| = n$. Since $0 < n$ we have that $X \neq \emptyset$. For every $x \in X$ the restriction of id_X to $\{x\}$ is injective, so

$$1 = |\{x\}| \leq |X| = n.$$

The Schröder-Bernstein Theorem now gives $n = 1$. $\Diamond$

6.15 Theorem *If* x *is a finite cardinal and* y *is an infinite cardinal then* $x < y$.

Proof We know that either $x \leq y$ or $y \leq x$. Now the latter implies, by 6.13, that y is finite, a contradiction. Hence we must have $x < y$. $\Diamond$

6.16 Theorem *For every* $n \in \mathbb{F}$ *with* $n \neq 0$ *there is a unique* $m \in \mathbb{F}$ *such that* $m + 1 = n$. *Moreover, if* $a < n$ *then* $a \leq m$.

Proof Since $\mathbb{F}$ is totally ordered under $\leq$, it follows from 6.14 that $1 \leq n$ and so, by 6.7 and 6.13, there exists $m \in \mathbb{F}$ such that $n = m + 1$. The uniqueness of m follows from 6.8.

Suppose now that $a < n = m + 1$. By 6.7 there exists $b \in \mathbb{F}$ such that $n = a+b$, and $b > 0$. By the first part of the theorem there exists c such that $b = c + 1$ and so

$$m + 1 = n = a + b = (a + c) + 1$$

whence, by 6.8, we have

$$m = a + c \geq a + 0 = a. \quad \Diamond$$

The unique m of 6.16 is often referred to as the *predecessor* of $n \in \mathbb{F}$ and is written $n - 1$. Note by 6.16 that for $n \neq 0$ we have

$$\{x \in \mathbb{F} \; ; \; n - 1 < x < n\} = \emptyset,$$

i.e. there is no cardinal lying strictly between $n-1$ and n. Also, writing $n + 1$ for n in 6.16 we obtain

$$(n + 1) - 1 = n,$$

so that n is the predecessor of $n + 1$ and there is no cardinal lying strictly between n and $n + 1$.

6.17 Theorem *If E is a subset of $\mathbb{F}$ such that*

(1) $0 \in E$;

(2) $n \in E$ *implies* $n + 1 \in E$,

then E coincides with $\mathbb{F}$.

Proof Suppose, by way of obtaining a contradiction, that we have $C_{\mathbb{F}}(E) \neq \emptyset$. Then since, by 6.2, the set $\mathbb{F}$ is well ordered the subset $C_{\mathbb{F}}(E)$ has a smallest element, n say. Now $n \neq 0$ by (1), and so by 6.16 there exists $m \in \mathbb{F}$ such that $n = m+1$. This implies that $m < n$ and hence, by the minimality of n, we have $m \notin C_{\mathbb{F}}(E)$. Thus $m \in E$ and by (2) we have the contradiction $n = m + 1 \in E$. We conclude therefore that $C_{\mathbb{F}}(E) = \emptyset$ and hence that $E = \mathbb{F}$. $\diamond$

We can now proceed to give an explicit description of the set $\mathbb{N}$ of natural numbers. Whatever our intuitive idea of this set may have been up until now, we at least agree that it is an example of the following type of set.

Definition By a *Peano set* we mean a set E on which there is defined a total order $\leq$ such that

(1) E has a smallest element α;

(2)   every $x \in E$ has a 'successor' $x^+ \in E$, in the sense that $x < x^+$ and if $y \in E$ is such that $x < y \leq x^+$ then $y = x^+$;

(3)   if X is a subset of E such that

(a) $\alpha \in X$;

(b) $x \in X$ implies $x^+ \in X$;

then X coincides with E.

As far as $\mathbb{N}$ is concerned, the total order is the natural order, the smallest element is 0, the successor of $n \in \mathbb{N}$ is $n + 1$, and the validity of (3) above is assured by the induction principle.

Taking another look at the set $\mathbb{F}$ of finite cardinals, we see by 6.2 that it too is totally ordered, by 6.14 it has a smallest element 0, and by 6.16 the successor of n is $n + 1$ (since the predecessor of $n + 1$ is n). Taking 6.17 into account, we see that $\mathbb{F}$ is also an example of a Peano set.

Our objective now is to prove that, in a sense to be made precise, all Peano sets 'look the same'; that, roughly speaking, there is only one such set. This being so, we can then agree to identify all Peano sets and thereby identify the set $\mathbb{N}$ of natural numbers with the set $\mathbb{F}$ of finite cardinals. Intuitively, this can be done by the association

$$0 \longleftrightarrow \mathbf{0} = |\emptyset|$$
$$1 \longleftrightarrow \mathbf{1} = |\{\emptyset\}|$$
$$2 \longleftrightarrow \mathbf{2} = |\{\emptyset, \{\emptyset\}\}|$$
$$3 \longleftrightarrow \mathbf{3} = |\{\emptyset, \{\emptyset\}, \{\emptyset, \{\emptyset\}\}\}|$$
$$\vdots$$

i.e. by considering the mapping $f : \mathbb{N} \to \mathbb{F}$ given by $f(n) = \mathbf{n}$ where $\mathbf{n}$ is the cardinal of a set with n elements. At the intuitive level this is fine, but it obscures the difficulty : we have to act without any proper knowledge of the set $\mathbb{N}$ of natural numbers. We proceed as follows, first defining precisely what we mean by saying that two ordered sets 'look the same'.

Definition Let E be ordered by $\leq_1$ and let F be ordered by $\leq_2$. Then a mapping $f : E \to F$ is called an *order isomorphism* if

(1) f is surjective;
(2) $(\forall x, y \in E)$ $x \leq_1 y \iff f(x) \leq_2 f(y)$.
When there exists an order isomorphism $f : E \to F$ we say that E and F are *order isomorphic*.

Note that an order isomorphism is necessarily a bijection since, by (2), we have

$$f(x) = f(y) \implies f(x) \leq_2 f(y) \text{ and } f(x) \geq_2 f(y)$$
$$\implies x \leq_1 y \text{ and } x \geq_1 y$$
$$\implies x = y,$$

so that (2) implies in particular that f is injective.

Intuitively, to say that E and F are order isomorphic means that E, F differ only in the notation for the elements and the orders.

6.18 Theorem *If E and F are sets that are totally ordered with respect to $\leq_1$ and $\leq_2$ respectively, then $f : E \to F$ is an order isomorphism if and only if*

(1) *f is surjective;*
(2) *$(\forall x, y \in E)$ $x <_1 y \implies f(x) <_2 f(y)$.*

Proof $\Rightarrow$: It suffices to note that if $x <_1 y$ then $f(x) \leq_2 f(y)$ with equality excluded since f is injective.

$\Leftarrow$: If f satisfies (1) and (2) then from (2) we obtain

$$x \leq_1 y \implies f(x) \leq_2 f(y).$$

Suppose now that $f(x) \leq_2 f(y)$. Then we must have $x \leq_1 y$; for otherwise, since E is totally ordered, we would have $y <_1 x$ and, by (2), the contradiction $f(y) <_2 f(x)$. Thus f is surjective and

$$x \leq_1 y \iff f(x) \leq_2 f(y),$$

and so is an order isomorphism. $\Diamond$

The key result in showing that any two Peano sets are order isomorphic is the following, in which

$$[0, \mathbf{n}] = \{\mathbf{x} \in \mathbb{F} \ ; \ 0 \leq \mathbf{x} \leq \mathbf{n}\}.$$

6.19 Theorem *Let E be a Peano set with smallest element α and successor function $x \mapsto x^+$. Then for every $\mathbf{n} \in \mathbb{F}$ there is a unique mapping $\vartheta_{\mathbf{n}} : [0, \mathbf{n}] \to E$ such that*

(1) $\vartheta_{\mathbf{n}}(0) = \alpha$;

(2) $(\forall \mathbf{r} < \mathbf{n}) \ \vartheta_{\mathbf{n}}(\mathbf{r} + 1) = [\vartheta_{\mathbf{n}}(\mathbf{r})]^+$.

Proof We establish the proof by using 6.17 (the 'induction principle in $\mathbb{F}$'). Let S be the set of all $\mathbf{n} \in \mathbb{F}$ for which there exists a unique mapping $\vartheta_{\mathbf{n}} : [0, \mathbf{n}] \to E$ satisfying (1) and (2). Then $0 \in S$ since $[0, 0] = \{0\}$ and only one such mapping is possible, namely ϑ_0 given by $\vartheta_0(0) = \alpha$. By way of applying 6.17, suppose that $\mathbf{n} \in S$ with associated mapping $\vartheta_{\mathbf{n}}$. Define $h : [0, \mathbf{n} + 1] \to E$ by

$$h(\mathbf{x}) = \begin{cases} \vartheta_{\mathbf{n}}(\mathbf{x}) & \text{if } \mathbf{x} \leq \mathbf{n}; \\ [\vartheta_{\mathbf{n}}(\mathbf{n})]^+ & \text{if } \mathbf{x} = \mathbf{n} + 1. \end{cases}$$

Since $\vartheta_{\mathbf{n}}$ satisfies (1) and (2) on $[0, \mathbf{n}]$, it is clear that h satisfies (1) and (2) on $[0, \mathbf{n} + 1]$. That h is unique follows from the fact that if $k : [0, \mathbf{n} + 1] \to E$ also satisfies (1) and (2) then the restriction k' of k to $[0, \mathbf{n}]$ satisfies (1) and (2) and so, by the uniqueness of $\vartheta_{\mathbf{n}}$, we have $k' = \vartheta_{\mathbf{n}}$. Thus k' and h coincide on $[0, \mathbf{n}]$. Since also

$$k(\mathbf{n} + 1) = [k'(\mathbf{n})]^+ = [h(\mathbf{n})]^+ = h(\mathbf{n} + 1),$$

it follows that $k = h$. We thus deduce that $\mathbf{n} + 1 \in S$. It now follows by 6.17 that $S = \mathbb{F}$, whence the result follows. $\Diamond$

6.20 Corollary *There is a unique mapping* $\varphi : \mathbb{F} \to E$ *such that*

(1') $\varphi(0) = \alpha$;
(2') $(\forall n > 0)$ $\varphi(n + 1) = [\varphi(n)]^+$.

Proof Consider the mapping $\varphi : \mathbb{F} \to E$ given by

$$\varphi(n) = \vartheta_n(n)$$

where ϑ_n is as in 6.19. Clearly,

$$\varphi(0) = \vartheta_0(0) = \alpha.$$

Now the restriction of ϑ_{n+1} to $[0, n]$ satisfies (1) and (2) on $[0, n]$ and so must coincide with ϑ_n. Consequently

$$\varphi(n + 1) = \vartheta_{n+1}(n + 1) = [\vartheta_{n+1}(n)]^+ = [\vartheta_n(n)]^+ = [\varphi(n)]^+,$$

and hence φ satisfies (1') and (2').

To show that φ is unique, suppose that $h : \mathbb{F} \to E$ also satisfies (1') and (2'). Then the restriction of h to $[0, n]$ satisfies (1) and (2) on $[0, n]$ and so coincides with ϑ_n. It follows that

$$h(n) = \vartheta_n(n) = \varphi(n)$$

for every n, and hence that $h = \varphi$. $\Diamond$

6.21 Theorem *Every Peano set is order isomorphic to* $\mathbb{F}$.

Proof Let E be a Peano set with smallest element α and successor function $x \mapsto x^+$. By 6.20, there is a unique $\varphi : \mathbb{F} \to E$ such that (1') and (2') hold. These equations show that $\alpha \in \text{Im}\,\varphi$, and that for every $x = \varphi(n) \in \text{Im}\,\varphi$ we have

$$x^+ = [\varphi(n)]^+ = \varphi(n + 1) \in \text{Im}\,\varphi.$$

It follows from property (3) in the definition of a Peano set that $\text{Im}\,\varphi = E$. Thus φ is surjective.

On the other hand, consider the subset

$$A = \{x \in \mathbb{F} \; ; \; y < x \Rightarrow \varphi(y) < \varphi(x)\}.$$

It is clear that $0 \in A$. Also, if $\mathbf{x} \in A$ then

$$\mathbf{y} < \mathbf{x} + 1 \Longrightarrow \mathbf{y} \leq \mathbf{x}$$
$$\Longrightarrow \varphi(\mathbf{y}) \leq \varphi(\mathbf{x}) < [\varphi(\mathbf{x})]^{+} = \varphi(\mathbf{x} + 1),$$

which shows that $\mathbf{x} + 1 \in A$. It thus follows by 6.17 that $A = \mathbb{F}$. Calling on 6.18, we now see that $\varphi : \mathbb{F} \to E$ is an order isomorphism. $\Diamond$

By virtue of 6.21 we can agree to identify all Peano sets. Our intuitive concept of the set $\mathbb{N}$ of natural numbers can now be made concrete by identifying this with the set of finite cardinals.

It follows from 6.2 that $\mathbb{F}$ (and so $\mathbb{N}$) is well ordered, a fact that we used in Chapter Four in order to establish the principle of induction. It is convenient at this point to establish the general connection between well ordering and induction.

6.22 Theorem *Let E be ordered by $\leq$. Then E is well ordered by $\leq$ if and only if*

(1) $\leq$ *is a total order;*
(2) E *has a smallest element α;*
(3)   *if F is a subset of E such that*
  (a) $\alpha \in F$;
  (b) $\{x \in E \, ; \, x < r\} \subseteq F$ *implies $r \in F$,*
then $F = E$.

Proof To show that the conditions are necessary, suppose that E is well ordered with respect to $\leq$. Then every non-empty subset of E has a smallest element. Applying this to any two-element subset $\{x, y\}$ then to E itself, we see that (1) and (2) hold. As for (3), suppose that the subset F satisfies (a) and (b). By way of obtaining a contradiction, suppose that $F \neq E$ and consider $C_E(F)$. This set has a smallest element, y say. Then $y \notin F$ and so, by (a), $y \neq \alpha$. By (b) there exists $z < y$ with $z \notin F$; for otherwise all $z < y$ belong to F and (b) gives the contradiction $y \in F$. But then $z \in C_E(F)$ with $z < y$, and this contradicts the fact that y is the smallest element of $C_E(F)$. We conclude that $F = E$ and so (3) also holds.

As for the necessity, suppose that (1), (2), (3) hold. To show that E is well ordered, it is equivalent to show that if F is a subset that has no smallest element then $F = \emptyset$. So let F be such a subset and consider $C_E(F)$. By (1) and (2), α cannot be in F and so $\alpha \in C_E(F)$. Also, if $\{x \; ; \; x < r\} \subseteq C_E(F)$ then we cannot have $r \in F$ (since r would then be the smallest element of F, a contradiction). We must then have $r \in C_E(F)$. It now follows by (3) that $C_E(F) = E$ whence $F = \emptyset$ as required. $\Diamond$

We now turn our attention to infinite cardinals. Here there are several surprises in store. First let us note that infinite cardinals do not behave algebraically like finite cardinals.

Example If $m, n, p \in \mathbb{N}$ then $p + m = p + n$ implies $m = n$. This is the *cancellation law for addition* in $\mathbb{N}$. It is no longer true if we replace p by an infinite cardinal. For example, if $\mathbf{a}$ is infinite then we have $\mathbf{a} + \mathbf{0} = \mathbf{a} = \mathbf{a} + \mathbf{1}$. If the cancellation law held then we would obtain the contradiction $\mathbf{0} = \mathbf{1}$.

Similarly, if $m, n, p \in \mathbb{N}$ with $p \neq 0$ then $mp = np$ implies $m = n$. This is the *cancellation law for multiplication* in $\mathbb{N}$. It is no longer true if we replace p by an infinite cardinal. For example, the mapping $f : \{0, 1\} \times \mathbb{N} \to \mathbb{N}$ given by

$$f(x, n) = \begin{cases} 2n & \text{if } x = 0; \\ 2n + 1 & \text{if } x = 1, \end{cases}$$

is a bijection. So $2|\mathbb{N}| = |\mathbb{N}|$, and if $|\mathbb{N}|$ were cancellable for multiplication we would have the contradiction $2 = 1$.

Let us now consider the cardinal $|\mathbb{N}|$. We have seen that the set $\mathbb{F}$ of finite cardinals is infinite, so $|\mathbb{N}| = |\mathbb{F}|$ is infinite. We sometimes write $|\mathbb{N}|$ as $\aleph_0$ ('aleph zero').

Definition A set E is said to be *denumerable* if $|E| = |\mathbb{N}|$.

Thus a set is denumerable if it is equipotent to the set $\mathbb{N}$ of natural numbers; or, put in an imprecise way that can be confusing, 'has the same number of elements' as $\mathbb{N}$.

6.23 Theorem *The set $\mathbb{Q}$ of rationals is denumerable.*

Proof Since $\mathbb{N} \subset \mathbb{Q}$ we have $|\mathbb{N}| \leq |\mathbb{Q}|$. Now every rational number can be written in its lowest terms as $(-1)^i p/q$ where $i \in \{1, 2\}, p \in \mathbb{N}, q \in \mathbb{N} \setminus \{0\}$. Consider the mapping $f : \mathbb{Q} \to \mathbb{N}$ given by

$$f\big((-1)^i p/q\big) = 2^i \cdot 3^p \cdot 5^q.$$

By unique factorisation in $\mathbb{N}$ we see that f is an injection. Thus $|\mathbb{Q}| \leq |\mathbb{N}|$ and the result follows by the Schröder-Bernstein Theorem. $\Diamond$

6.24 Corollary *The set $\mathbb{Z}$ of integers is denumerable.*

Proof Since $\mathbb{N} \subset \mathbb{Z} \subset \mathbb{Q}$ we have $|\mathbb{N}| \leq |\mathbb{Z}| \leq |\mathbb{Q}| = |\mathbb{N}|$ so the result follows by the Schröder-Bernstein Theorem. $\Diamond$

A further surprise is the following.

6.25 Theorem $\mathbb{N} \times \mathbb{N}$ *is denumerable.*

Proof Consider the mapping

$$f : \mathbb{N} \times \mathbb{N} \to \mathbb{N}$$

given by $f(m, n) = 2^m \cdot 3^n$. By unique factorization, this is an injection. But the mapping

$$g : \mathbb{N} \to \mathbb{N} \times \mathbb{N}$$

given by $g(n) = (n, n)$ is also an injection. The result therefore follows by the Schröder-Bernstein Theorem. $\Diamond$

6.26 Corollary *If E and F are denumerable, so is $E \times F$.*

Proof From bijections $f : E \to \mathbb{N}$ and $g : F \to \mathbb{N}$ we can construct a bijection

$$(x, y) \longmapsto \big(f(x), g(y)\big)$$

of $E \times F$ to $\mathbb{N} \times \mathbb{N}$. $\Diamond$

6.27 Theorem *Every infinite set contains a denumerable subset.*

Proof Let E be an infinite set. By induction, for every $n \in \mathbb{N}$ we can construct as follows a subset E_n of E having n elements. In fact, for $n = 0$ the subset in question is $E_0 = \emptyset$; and if E_n is a subset having n elements then we can choose $x_n \in C_E(E_n)$ and form the subset $E_{n+1} = E_n \cup \{x_n\}$ which has $n + 1$ elements. By the construction of the E_n, the set $\{x_n \; ; \; n \in \mathbb{N}\}$ is then denumerable. $\Diamond$

6.28 Corollary *Every infinite subset of $\mathbb{N}$ is denumerable.*

Proof Let E be an infinite subset of $\mathbb{N}$. By 6.27, E contains a denumerable subset F. Then

$$|\mathbb{N}| = |F| \leq |E| \leq |\mathbb{N}|$$

and so, by the Schröder-Bernstein Theorem, $|E| = |\mathbb{N}|$. $\Diamond$

6.29 Corollary $|\mathbb{N}|$ *is the smallest infinite cardinal.*

Proof If E is an infinite set then there is a subset F of E that is denumerable. Since there is an injection from $\mathbb{N}$ into F, there is an injection $n \mapsto x_n$ of $\mathbb{N}$ into E and therefore $|\mathbb{N}| \leq |E|$. $\diamond$

So far, we have not been able to progress beyond $|\mathbb{N}|$. The next result opens the door to an infinity of infinities.

6.30 Theorem *For every set A we have $|A| < |P(A)|$.*

Proof The injection $x \mapsto \{x\}$ shows that $|A| \leq |P(A)|$. Now if we had equality then there would exist an injection $f : P(A) \to A$ and then, by 2.3, a mapping $g : A \to P(A)$ such that $g \circ f$ is the identity on $P(A)$. By 2.4, g is surjective. Consider now the element B of $P(A)$ given by

$$B = \{x \in A \; ; \; x \notin g(x)\}.$$

Since g is surjective there exists $y \in A$ such that $g(y) = B$. Now if $y \in B$ we have the contradiction $y \notin g(y) = B$; and if $y \notin B$ we have the contradiction $y \in g(y) = B$. We conclude, therefore, that no such injection f can exist and that consequently $|A| < |P(A)|$. $\diamond$

It follows from 6.30 that if we write

$$|\mathbb{N}| = \aleph_0, \quad |P(\mathbb{N})| = \aleph_1, \quad |P(P(\mathbb{N}))| = \aleph_2, \quad \ldots,$$

then we have the infinite chain of cardinals

$$0 < 1 < 2 < \cdots < \aleph_0 < \aleph_1 < \aleph_2 < \cdots.$$

Definition A set E is said to be *countable* if it is either finite or denumerable.

6.31 Theorem *If $E \neq \emptyset$ then the following statements are equivalent :*

(1) *E is countable;*
(2) *there is an injection $f : E \to \mathbb{N}$;*
(3) *there is a surjection $g : \mathbb{N} \to E$.*

Proof (1) $\Rightarrow$ (2) : If (1) holds then (2) is clear if E is denumerable. If, on the other hand, E is finite then by 6.15 we have $|E| < |\mathbb{N}|$ and so there is an injection $f : E \to \mathbb{N}$.

(2) $\Rightarrow$ (1) : Let $f : E \to \mathbb{N}$ be an injection. If E is finite then (1) holds trivially. If E is infinite then Im f, being equipotent to E, is an infinite subset of $\mathbb{N}$ and so, by 6.28, is denumerable. Thus we have

$$|E| = |\operatorname{Im} f| = |\mathbb{N}|$$

and so E is denumerable.

(2) $\Longleftrightarrow$ (3) : This is immediate from 2.3 and 2.4. $\Diamond$

6.32 Theorem *Let* $(E_i)_{i \in I}$ *be a family of sets each of which is countable. Then if the index set* I *is countable so also is* $\bigcup_{i \in I} E_i$.

Proof Since I is countable there is, by 6.31, an injection $f : I \to \mathbb{N}$. There is no loss in generality, therefore, if we assume that I is a subset of $\mathbb{N}$. The obvious injections

$$\mathbb{N} \longrightarrow \{0\} \times \mathbb{N} \longrightarrow I \times \mathbb{N} \longrightarrow \mathbb{N} \times \mathbb{N}$$

together with the Schröder-Bernstein Theorem and 6.31 now show that $I \times \mathbb{N}$ is denumerable. Thus there is a bijection

$$\varphi : \mathbb{N} \to I \times \mathbb{N}.$$

Since each E_i is countable there is, for every $i \in I$, a surjection $g_i : \mathbb{N} \to E_i$. Consider the mapping

$$\vartheta : I \times \mathbb{N} \to \bigcup_{i \in I} E_i$$

given by

$$\vartheta(i, n) = g_i(n).$$

Since each g_i is surjective, so is ϑ. The composite mapping

$$\mathbb{N} \xrightarrow{\ \varphi\ } I \times \mathbb{N} \xrightarrow{\ \vartheta\ } \bigcup_{i \in I} E_i$$

is then a surjective mapping from $\mathbb{N}$ onto $\bigcup_{i \in I} E_i$. The result now follows by 6.31. $\Diamond$

We now give an example of a set that is not countable. For this purpose, we recall that every real number $r \in [0, 1[$ has a unique decimal representation

$$r = 0 \cdot r_0 r_1 r_2 \ldots,$$

this being an abbreviation for the equality

$$r = \frac{r_0}{10} + \frac{r_1}{10^2} + \frac{r_2}{10^3} + \cdots$$

in which each r_i is such that $0 \leq r_i \leq 9$ and not all r_i from some point on are equal to 9. This condition is necessary in order to have uniqueness. For example, if

$$r = 0 \cdot 9999 \ldots$$

then

$$10r = 9 \cdot 9999 \ldots$$

so $9r = 10r - r = 9$ and hence $r = 1$.

6.33 Theorem *The half-open interval $[0, 1[$ of real numbers is uncountable.*

Proof Suppose, by way of obtaining a contradiction, that $[0, 1[$ were countable. We know that this interval is infinite. It is therefore denumerable, so there is a bijection $f : [0, 1[\to \mathbb{N}$; i.e. we can list the elements of $[0, 1[$ as a sequence $(\alpha_i)_{i \geq 0}$ where α_i has the decimal representation

$$\alpha_i = 0 \cdot \alpha_{i0} \alpha_{i1} \alpha_{i2} \ldots.$$

Consider now the real number $x \in [0, 1[$ whose decimal representation is

$$x = 0 \cdot x_0 x_1 x_2 \ldots$$

where, for every $i \geq 0$, $x_i \neq \alpha_{ii}$ and $x_i \neq 9$ for any i; i.e. x is such that its first digit is different from the first digit of α_0, its second digit is different from the second digit of α_1, and so on. Then x has a decimal representation that differs in at least one place from every decimal α_i in $[0, 1[$ and we have the contradiction $x \notin [0, 1[$. Thus $[0, 1[$ is not denumerable and so is uncountable. $\diamond$

6.34 Corollary $\mathbb{R}$ *is uncountable.*

Proof By 6.29 and 6.23 we have $|\mathbb{N}| < \|0,1\| \leq |\mathbb{R}|.$ $\diamond$

6.35 Theorem $\|0,1\| = \|0,1\| = |\mathbb{R}| = |\mathbf{P}(\mathbb{N})| = \aleph_1.$

Proof Just as every real number $r \in [0,1[$ has a unique decimal representation, so does every $r \in [0,1[$ have a unique binary representation

$$r = 0 : r_0 r_1 r_2 \ldots,$$

which is an abbreviation for the equality

$$r = \frac{r_0}{2} + \frac{r_1}{2^2} + \frac{r_2}{2^3} + \cdots$$

in which each $r_i \in \{0,1\}$ with not all r_i from some point on equal to 1.

We can define a mapping $f : [0,1[\to \mathbf{P}(\mathbb{N})$ by

$$f(r) = \{i \in \mathbb{N} ;\ r_i = 1\}.$$

By the uniqueness of binary expansions, f is injective and so

$$\|0,1\| \leq |\mathbf{P}(\mathbb{N})|.$$

But we can also define a mapping $g : \mathbf{P}(\mathbb{N}) \to [0,1[$ by

$$g(X) = 0 : x_0 0 x_1 0 x_2 0 \ldots$$

where

$$x_i = \begin{cases} 0 & \text{if } i \notin X; \\ 1 & \text{if } i \in X. \end{cases}$$

Clearly, g is also an injection and so

$$|\mathbf{P}(\mathbb{N})| \leq \|0,1\|.$$

By the Schröder-Bernstein Theorem we then have

$$\|0,1\| = |\mathbf{P}(\mathbb{N})|.$$

Finally, the equality

$$\|0,1\| = |\mathbb{R}|$$

follows, for example, from the fact that $\|0,1\| = \,]0,1\|$ and that the mapping $f :\,]0,1[\to \mathbb{R}$ given by

$$f(x) = \frac{1-2x}{x(1-x)}$$

is a bijection. $\diamond$

Example We have seen in 6.25 that $|\mathbb{N} \times \mathbb{N}| = |\mathbb{N}|$. It follows that $\aleph_0^2 = \aleph_0$. It is also true that $\aleph_1^2 = \aleph_1$. To see this, consider the half-open unit square

$$I = [0, 1[\times [0, 1[.$$

For every $(x, y) \in I$ we have unique decimal expansions

$$x = 0 \cdot x_0 x_1 x_2 \ldots,$$
$$y = 0 \cdot y_0 y_1 y_2 \ldots.$$

Consider the mapping

$$f : [0, 1[\times [0, 1[\to [0, 1[$$

given by

$$f(x, y) = 0 \cdot x_0 y_0 x_1 y_1 x_2 y_2 \ldots.$$

Clearly, this is an injection and so

$$\|[0, 1[\times [0, 1[\| = \|[0, 1[\|,$$

i.e. we have $\aleph_1^2 \leq \aleph_1$. But $\aleph_1 = 1 \cdot \aleph_1 \leq \aleph_1^2$ and so $\aleph_1^2 = \aleph_1$. In fact, although it is more difficult to do so, it can be shown that $\mathbf{a}^2 = \mathbf{a}$ for every infinite cardinal $\mathbf{a}$.

Example The cardinals $\aleph_0, \aleph_1$ are such that

$$\aleph_0 \aleph_1 = \aleph_0 + \aleph_1 = \aleph_1.$$

To see this, observe by 6.6 that

$$\aleph_0 < \aleph_1 \implies \aleph_0 \aleph_1 \leq \aleph_1^2 = \aleph_1;$$
$$1 \leq \aleph_0 \implies \aleph_1 = 1\aleph_1 \leq \aleph_0 \aleph_1,$$

and so $\aleph_0 \aleph_1 = \aleph_1$.

Again using 6.6 we have

$$\aleph_1 = \aleph_1 + 0 \leq \aleph_1 + \aleph_0$$
$$\leq \aleph_1 + \aleph_1$$
$$= |(\mathbb{R} \times \{1\}) \cup (\mathbb{R} \times \{2\})|$$
$$= |\mathbb{R} \times \{1, 2\}|$$
$$= 2\aleph_1$$
$$\leq \aleph_0 \aleph_1$$
$$= \aleph_1.$$

Consequently $\aleph_0 + \aleph_1 = \aleph_1$.

In fact, though it is more difficult to do so, it can be shown that if a, b are non-zero cardinals at least one of which is infinite then $ab = a + b = \max\{a, b\}$. The arithmetic of infinite cardinals is therefore very much simpler than that of finite cardinals.

In the chain of cardinals we have seen that $\aleph_0 < \aleph_1$. A celebrated problem that was first raised by Cantor and has since intrigued mathematicians is the following : *does there exist a cardinal lying strictly between $\aleph_0$ and $\aleph_1$?* The assertion that there is not, which is equivalent to the assertion that every uncountable set contains a subset equipotent to $\mathbb{R}$, is known as the *continuum hypothesis*. It was shown as recently as 1963 by P. J. Cohen that this question is 'unanswerable' in the sense that the hypothesis is independent of the usual axioms of set theory. What this amounts to is that a mathematician may choose to accept or reject the hypothesis depending on the needs of the mathematics he wishes to develop.

Index